Sand Castles and Mud huts

and

COMPREHENSION ACTIVITIES FOR A LEVEL CHEMISTRY

Jeffrey Hancock

King Edward's School
Birmingham

Hodder & Stoughton

A MEMBER OF THE HODDER HEADLINE GROUP

GW01460155

British Library Cataloguing in Publication Data

Hancock, D.J.
    Sand castles and mud huts: Comprehension
    activities for A level chemistry.
    I. Title
    540

    ISBN 0 340 543698

First published by Hodder and Stoughton Educational 1991
Impression number   10 9 8 7 6 5 4 3
Year   1999 1998 1997 1996 1995

Typeset by Keyset Composition, Colchester, Essex
Printed in Great Britain for Hodder & Stoughton Educational, a division of Hodder
Headline Plc, 338 Euston Road, London NW1 3BH by The Looseleaf Company, Melksham, Wilts.

Contents

Acknowledgements

I am most grateful to the following people and organisations who gave me an immense amount of help, often taking a great deal of trouble and time to do so.

Mrs S L Abbott, Information Pharmacist, Windsor Pharmaceuticals Ltd. Dr Diane Bannister, Technical Information Officer, Boots Pharmaceuticals. Mr S R Blunt and Dr R K Bramley, The Forensic Science Laboratory, Birmingham. Mr M Checkley, Principal Scientific Officer, Coventry City Council Environmental Services Department. Mr A N Crawford, District Dental Officer, South Manchester Health Authority. Mr H L Curley, Production Director, T & R Theakston Ltd, Brewers. Ms M Dando, Environmental Scientist, Enviro Technology Services. Mr V H Hyde, Technical Resources Manager, L'Oreal Golden Ltd. Mr Geraint Jenkins, Dental Programmes Executive, The Wrigley Company Ltd. PowerGen plc. Miss S M Thompson, Consultant Ophthalmic Surgeon, Sandwell and Dudley Health Authority. Mr J Weiner, Quality Assurance Chemist, Courage Ltd, Brewers. Dr K L Woods, Consultant Physician, The Leicester Royal Infirmary.

Numerous students at King Edward's and my colleagues in the chemistry department at King Edward's: Derek Benson, Peter Russell, Robin Smith and Rob Symonds.

And most especially, my wife, Miriam, for help in every way imaginable.

Introduction

Advanced Level Chemistry presents teachers with something of a dilemma. Some of our students will go on to study science at a higher level, and we must give them the necessary grounding for that. On the other hand, many or most of them will not, and studying shapes of molecules, organic mechanisms or reaction kinetics can seem dull and academic to them. Where is the *real* world?

This pack is an attempt to meet that problem. Each section (bar two) begins with a short passage about a real world problem: how anaesthetics work, preventing sun-burn, smoke detectors, why sweet things taste sweet, and so on. This is then followed by some questions. Some of these questions are GCSE in type, asking the student to explain a bit of the passage in his or her own words, to weigh up some evidence, analyse data or judge opposing arguments. But the majority of the questions are designed to apply academic chemistry to these real world problems.

I hate those problem books which the student can only use when he or she has finished the whole A level course! So I have tried very hard to restrict the questions on any passage to *one traditional chemistry topic*. So, for example, the questions on the passage 'Tonight, Josephine!' are (almost) all about the traditional chemistry of halogenoalkanes, and once you have taught this, the student should be able to make some sort of attempt at them. Similarly, 'Taking drugs' is only about reaction kinetics, and so on. (It's perhaps fair to add that all of the topics use GCSE chemistry, too.)

Some of the material may look hard. But if you can persuade your students not to be put off, they should find that most of it is actually quite interesting, and not as difficult as it looked at first. I've tried to steer them in the right direction: the earlier questions are usually more straightforward traditional material, leading on to the related real world stuff. Often the answer is in the text; much of this material can be used for comprehension practice.

A few of the questions *are* hard, however. Chemistry is about reality, and reality is complicated! Sometimes a question may have crept in that is not strictly related to the central topic: these are marked with an asterisk. I have used traditional or commercial names for many compounds, especially if the IUPAC names are complex.

I have tried to check everything against the relevant literature. Please tell me of any errors.

Sand castles and mud huts © 1991 Jeffrey Hancock, published by Hodder and Stoughton Educational

1. *Hindenburg revisited?*

Two points to think about. First, all our fossil fuels will run out eventually. And second, all of them pollute the Earth.

Exactly how long we've got before the pumps run dry we cannot say: it depends on rates of usage and discovery. But we can estimate when fuel production will reach its peak and start to decline. These estimates range from the year 2020 to 2050. The problem is urgent: how old will your children be then?

Even if we get rid of the SO_2, the smoke, the CO and the nitrogen oxides, there is still the CO_2 and the greenhouse problem. Everything pollutes!

Well – no, it doesn't, actually. If we were to burn hydrogen as our fuel, the product would just be water, hardly a pollutant.

How could we use hydrogen?

Not to make electricity. By the time we have run out of coal (still the major fuel for power stations), we shall probably do things differently. There may be more nuclear power, despite the unsolved problems of waste disposal. There will certainly be much more economical use of fuel. (In Sweden the houses are triple or quadruple glazed. Has yours even got double glazing?) There will also be more use of renewables, like wind power. (20% of Denmark's electricity will come from wind power by the year 2000.)

Could hydrogen be used to replace hydrocarbon fuels in diesel, petrol and aircraft engines?

Wait a minute. Hydrogen burns. Hydrogen *explodes*. Should we really put it in aircraft, with *people*? You have probably heard of the Hindenburg, a balloon (commonly called an airship, but actually filled with hydrogen) that went into passenger service between Germany and the USA in the summer of 1936. In May 1937 it caught fire as it came in to land at Lakehurst, New Jersey. It was burnt out in minutes. But of the 97 people on board, only 35 died. Compare that with the fire on Boeing 737 G-BGJL at Manchester airport on 22 August 1985. The aircraft didn't leave the ground, but 55 of the 137 passengers and crew were killed. Sadly, there are dangers involved in the use of any fuel, but perhaps the dangers of using hydrogen aren't much different.

There are two further problems, however.

The first is getting the stuff. Hydrogen is different from hydrocarbon fuels. You can't just extract it from underground; it has to be manufactured. At present this is done by one of two methods: by electrolysis of water (which requires cheap electricity) or by the reaction of coal or methane with steam. A promising alternative is the direct splitting of water by some reversible chemical reaction. Several thousand different chemical cycles have been suggested. One possibility is:

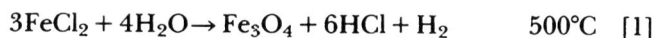

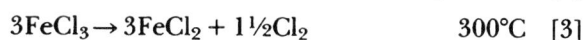

$$3FeCl_2 + 4H_2O \rightarrow Fe_3O_4 + 6HCl + H_2 \qquad 500°C \quad [1]$$

$$Fe_3O_4 + 1\tfrac{1}{2}Cl_2 + 6HCl \rightarrow 3FeCl_3 + 3H_2O + \tfrac{1}{2}O_2 \quad 100°C \quad [2]$$

$$3FeCl_3 \rightarrow 3FeCl_2 + 1\tfrac{1}{2}Cl_2 \qquad 300°C \quad [3]$$

Alternatively, could we use the energy of sunlight to do it directly? After all, plants absorb sunlight and carry out photosynthesis:

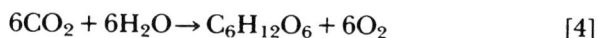

$$6CO_2 + 6H_2O \rightarrow C_6H_{12}O_6 + 6O_2 \qquad [4]$$

and this requires 2803 kJ per mole of glucose produced. The search is

CONTINUED

now on for a suitable compound to absorb the sunlight. It would normally be an oxidising agent, capable of oxidising water to oxygen (equation 5). Then when it absorbed light energy (equation 6), it would become a reducing agent, able to convert water to hydrogen (equation 7). In other words, we are looking for a compound X, with a reduced form Y, such that:

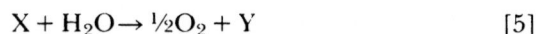

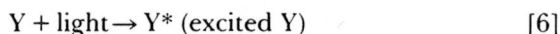

$$X + H_2O \rightarrow \tfrac{1}{2}O_2 + Y \qquad [5]$$

$$Y + light \rightarrow Y^* \text{ (excited Y)} \qquad [6]$$

$$Y^* + H_2O \rightarrow H_2 + X \qquad [7]$$

The second problem attached to the use of hydrogen lies in storing it. The boiling point of the liquid is 20K, $-253°C$. This will obviously require special handling techniques: you won't just drive up to a pump and fill a tank from a hose. Fortunately some transition metal alloys can absorb vast quantities of hydrogen gas, much as a sponge absorbs water. For example, there is an iron-titanium alloy that can form $FeTiH_x$, where x can take any value up to 1.95. The alloy $LaNi_5$ can form $LaNi_5H_6$, which actually contains more hydrogen per unit volume than liquid H_2 itself! Instead of a petrol tank then, a car would use a block of this alloy sponge. To fill it with hydrogen, it would be connected to a high pressure hydrogen tank. Of course, this kind of fuel tank will burn, but only very gently. Have you ever seen a petrol tank on fire? Yet we happily drive around with gallons of the stuff slopping about under the rear seat passengers! A hydrogen fuel tank would actually be safer.

Questions

1. (a) State the first law of thermodynamics.

   (b) What is Hess's law?

   (c) Explain how Hess's law is related to the first law of thermodynamics.

2. (a) Define enthalpy of formation.

   (b) What else must be specified if the *standard* enthalpy of formation is to be defined?

   (c) What is meant by enthalpy of combustion?

3. (a) Write down equations (including state symbols) for reactions whose enthalpy changes would be:

   (i) $\Delta H_f^{\circ}(H_2O)$ (ii) $\Delta H_c^{\circ}(H_2)$
   (iii) $\Delta H_f^{\circ}(CO_2)$ (iv) $\Delta H_c^{\circ}(CO)$.

   (b) How are the following related to each other?

   (i) $\Delta H_c^{\circ}(H_2)$ and $\Delta H_f^{\circ}(H_2O)$

   (ii) $\Delta H_c^{\circ}(CO)$, $\Delta H_f^{\circ}(CO)$ and $\Delta H_f^{\circ}(CO_2)$

   (iii) $\Delta H_f^{\circ}(CH_4)$, $\Delta H_f^{\circ}(H_2O)$, $\Delta H_c^{\circ}(C)$ and $\Delta H_c^{\circ}(CH_4)$.

4. (a) At present one method of manufacturing H_2 is by the reaction of methane and steam:

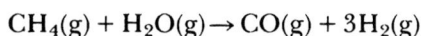

   $$CH_4(g) + H_2O(g) \rightarrow CO(g) + 3H_2(g)$$
   then $$CO(g) + H_2O(g) \rightarrow CO_2(g) + H_2(g)$$

Given the following standard enthalpies of formation, calculate a value for the standard enthalpy change for each of the reactions above.

	$CH_4(g)$	$H_2O(g)$	$CO(g)$	$CO_2(g)$
$\Delta H_f^{\circ}/kJ\ mol^{-1}$	-74.8	-241.8	-110.5	-393.5

CONTINUED

Sand castles and mud huts © 1991 Jeffrey Hancock, published by Hodder and Stoughton Educational

(b) The overall process (both reactions combined) is endothermic. So why is it useful at present?

(c) Why won't it be useful in the future?

Questions 5–6 examine the usefulness of various fuels. We are going to look at methane (the major component of natural gas), hydrogen, both gaseous and liquid, and petrol. Petrol is a complex mixture, so we'll use just one, fairly typical, component: octane, C_8H_{18}.

TABLE 1

Fuel	Standard enthalpy of combustion /kJ mol^{-1}	Density of liquid/g cm^{-3}
CH_4	−890.3	−
H_2	−285.8	0.071
C_8H_{18}	−5470.2	0.703

5. The amount of heat released when a fuel burns is important, but it's not the only factor. If it is to be used in an aircraft, the mass of the fuel is important too. A heavy fuel means a smaller payload.

(a) Calculate the heat produced per gram of each fuel in table 1, that is, the specific enthalpy. (Note that it is the heat *produced*, so it has no minus sign.) (Relative atomic masses H = 1, C = 12.)

(b) Which is the best fuel from the point of view of specific enthalpy?

6. The *volume* occupied by the fuel is also important, for if it takes up a lot of room, the fuel tanks will have to be huge. The enthalpy density is the heat produced per cubic decimetre of the fuel.

(a) Use the data in table 1 to calculate the enthalpy density for gaseous CH_4 and H_2 (at 298K, 1 atmosphere pressure) and for liquid H_2 and C_8H_{18}. (1 mole of any gas at 298K and 1 atmosphere pressure occupies about 24 dm^3.)

(b) Which fuel has the best enthalpy density?

7. In the medium term, the major problem with carbon-containing fuels is the fact that they produce CO_2, inevitably warming the Earth, with potentially catastrophic consequences. Are they all equally damaging? Compare methane and octane in terms of joules produced per mole of CO_2.

8. (a) Calculate an enthalpy change for reaction [1] in the passage, given the following enthalpies of formation (kJ mol^{-1}):

$FeCl_2(s)$	$H_2O(g)$	$Fe_3O_4(s)$	$HCl(g)$
−341.8	−241.8	−1118.4	−92.3

(b) The enthalpy change is −251.7 kJ mol^{-1} for reaction [2], and +173.1 kJ mol^{-1} for reaction [3]. Calculate the *total* enthalpy change for all three reactions.

(c) Compare your answer with the data given in (a), and comment.

9. (a) Use a data book to find how much energy is needed for the process:

$$H_2O(l) \rightarrow H_2(g) + \frac{1}{2}O_2(g)$$

(b) What is the minimum possible difference in energy between X and the reduced form of X after excitation by light (i.e. between X and Y*)?

(c) What is the minimum amount of energy that the light must supply (if the process is perfectly efficient)?

2. Fire detectors

There are two common types of household smoke detector.

Photoelectric detectors depend on the fact that smoke particles scatter light. (Next time you go to the cinema, look up at the light from the projector. If smoke gets into the beam, the light is scattered and the beam becomes visible.) The fire detector chamber also has a light in it, and a light-sensitive detector, tucked out of the way of the light beam. When smoke gets into the chamber, it scatters the light. The detector now 'sees' the light for the first time, and sounds an alarm.

The other type, perhaps the most common, is widely available in do-it-yourself shops. It uses a radioactive source, the 241 isotope of element number 95, americium. The radiation emitted from it ionises the air inside the device. Because this produces ions in the air, it turns it into an electrical conductor. A small electrical current is constantly passed through this ionised air. When the products of combustion enter the detector, they make the gas a less efficient conductor, and so a smaller current flows through the device. This change in current triggers the alarm.

Questions

On radioactivity

1. (a) Explain the words:

   (i) isotope, (ii) radioactive.

   (b) Element number 83 (bismuth) is the last element in the periodic table to have a stable isotope. All the later elements have only radioactive ones. Why are the heavy elements so unstable?

2. You are given the data in table 1.

TABLE 1

Nuclide	Half life /years	Decay	Energy of emission /MeV
^{210}Pb	22.7	β	0.061
^{228}Th	1.91	α	5.521
^{241}Am	458	α	5.640

   (a) What is meant by 'half life'?

   (b) What are α and β particles?

   (c) Why is the energy of the emitted particle important?

   (d) Suggest why:

   (i) ^{241}Am is used in preference to ^{210}Pb,

   (ii) ^{241}Am is used in preference to ^{228}Th.

CONTINUED

Sand castles and mud huts © 1991 Jeffrey Hancock, published by Hodder and Stoughton Educational

3. (a) ^{241}Am decays by α emission. Write an equation for the decay.

(b) Then there are another 7 α and 4 β emissions. What is the final nuclide produced in the device?

(c) ^{241}Am was originally made by bombarding ^{238}U with α particles. This produced an intermediate nuclide, which then formed the ^{241}Am by emitting a β particle (and nothing else). Write equations for the two processes.

4. Is there a risk attached to having a radioactive material in the house? How can any risk be minimised?

5. (a) Nuclear physicists use the unit of energy 'million electron volts' (MeV). Use a data book to find a suitable conversion factor and calculate the energy of the α particle emitted by ^{241}Am in kJ mol^{-1} of α particles.

(b) In fact, not all the ^{241}Am decays by α emission. In 81% of cases, there is emission of an α particle of energy 5.430 MeV, followed by emission of γ radiation. Why is the γ radiation emitted? What is its energy (in MeV)?

6. 1.0×10^{-6} g of ^{241}Am were sealed into a smoke detector in 1984. What mass of the nuclide remained when archaeologists excavated the house in AD 3816?

7. (a) Explain how the radiation emitted from the ^{241}Am ionises the air molecules, and how this enables the gas to pass a small electric current.

(b) When a house catches fire, smoke particles arrive at the detector, and reduce the current flowing across it. Presumably the smoke particles are ionised in just the same way as the air molecules. Why, then, does the current fall?

(c) Does the situation alter if the fire produces no smoke at all, but only CO_2 and water? Explain.

8. If both sorts of smoke detector are available, which should we choose? The most important consideration is obviously their effectiveness, but that doesn't help much: tests show that there is little difference.

(a) How about maintenance? Photoelectric devices get less sensitive, as dirt builds up inside them, so they have to be cleaned regularly. Explain how dirt affects an ionisation detector.

(b) How about the lifetime of the device? An ionisation detector sets off the alarm because the current passing across the device has fallen – by 5% for example. But the current will inevitably fall as time passes, because the radioactivity of the ^{241}Am decreases. How long will a device of this type last? How long will it be before the radioactivity of the ^{241}Am falls by 5%?

3. Gut contents, male suicide

Gas chromatography and mass spectroscopy

One morning in July 1991, a middle-aged man was found dead in bed in his house in the Midlands. It was eventually established that he was the managing director of a local firm. He had returned unexpectedly from a business trip abroad to find that his wife had left him. Had he killed himself?

In all cases of sudden death, a post-mortem examination is carried out. As a part of it, the contents of the man's gut were sent for analysis.

Where do you start? If you suspect one particular compound, you can do a test for it, but a thousand or more different substances might be present.

In this case, two linked techniques were used. The first was gas chromatography. You have probably seen and used paper chromatography, perhaps to separate the components of an ink or to distinguish different food dyes. Gas chromatography operates on a similar principle (fig 1). Instead of paper, the stationary phase is a liquid (adsorbed on to an inert solid and packed into a long column). Instead of the solvent used in paper chromatography, the moving phase is a gas, which is blown through the column at a steady rate. The mixture to be analysed is injected into the beginning of the column and it vaporises (because the column is contained in an oven at a high temperature). This vapour is then swept through the column by the carrier gas. In paper chromatography the different spots move at different speeds along the paper. In gas chromatography, the different compounds in the mixture move through the column at different speeds, so they reach the end of the column at different times. The various compounds in the dead man's stomach could thus be separated.

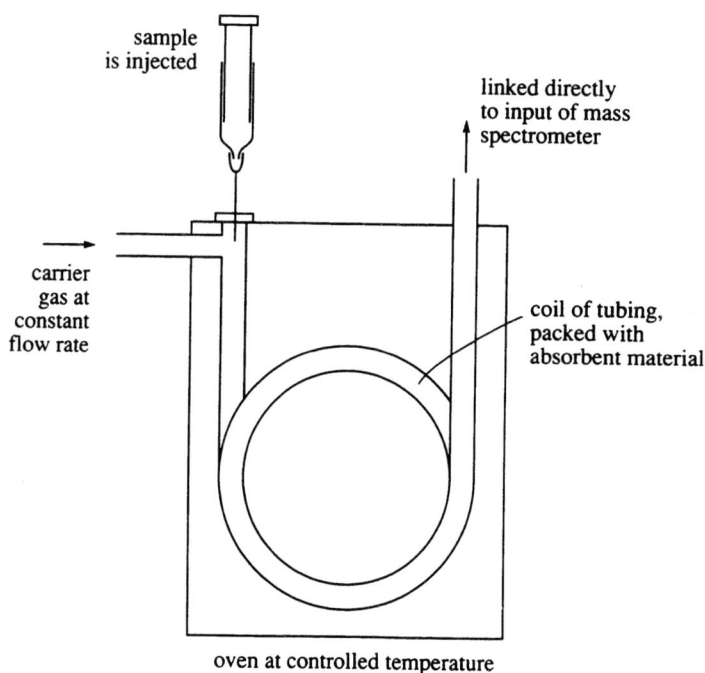

FIG 1   The gas chromatograph

CONTINUED

Sand castles and mud huts © 1991 Jeffrey Hancock, published by Hodder and Stoughton Educational

Each compound emerging from the gas chromatograph can be examined by mass spectrometry, often using a mass spectrometer directly linked to the output of the gas chromatograph.

The principle of the mass spectrometer is straightforward (fig 2). The molecules of the substance are bombarded with high energy electrons, which knock one or more electrons out of the molecules and thus ionise them:

$$M + e \rightarrow M^+ + 2e$$

These positive ions are accelerated by charged plates and then deflected by a magnetic field. If the deflection is exactly right, the ions arrive at a detector and are recorded. Variation in the speed of the ions or in the magnetic field strength allows all the possible ions to be detected, and if the instrument is calibrated, the mass of the ion can be determined.

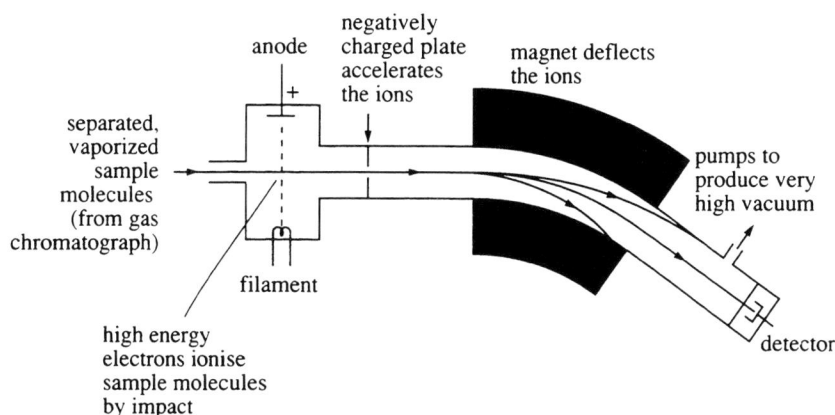

FIG 2   The mass spectrometer

Questions

1. (a) Look at table 1. What are isotopes?

(b) If you look up the relative atomic mass of naturally occurring bromine in a data book, you will find that it is just less than 80. Why is this?

(c) Calculate the accurate relative atomic mass of bromine.

TABLE 1   Accurate relative atomic masses

	Isotope				
	^{12}C	^{14}N	^{16}O	^{79}Br	^{81}Br
Isotopic mass	12.0000	14.0031	15.9949	78.9183	80.9163
Abundance/%	98.90	99.63	99.76	50.69	49.31

2. In a recent investigation, a sample of gas from the lungs of a dead woman was analysed. The M^+ peak had a mass of 28. Was this due to nitrogen only or could she have died from CO poisoning? High resolution mass spectrometry gave an accurate relative molecular mass of 27.9949.

(a) Which gas was it?

(b) *Briefly* summarise the evidence you would present to the inquest. (Remember that not everyone will have a scientific training.)

CONTINUED

The simplest possible ion is produced in the mass spectrometer by removal of one electron from the molecule. This ion (the molecule-ion, M^+) has so much energy as a result of the electron impact that it breaks up further to give other smaller ions. So any organic molecule will produce many positive ions. These can be detected by the spectrometer and displayed in a mass spectrum which records the mass of each ion on the x-axis, and the ion's abundance on the y-axis. The *most abundant* ion will be set to 100%.

Four mass spectra obtained from the contents of the dead man's stomach are shown in figs 3–6.

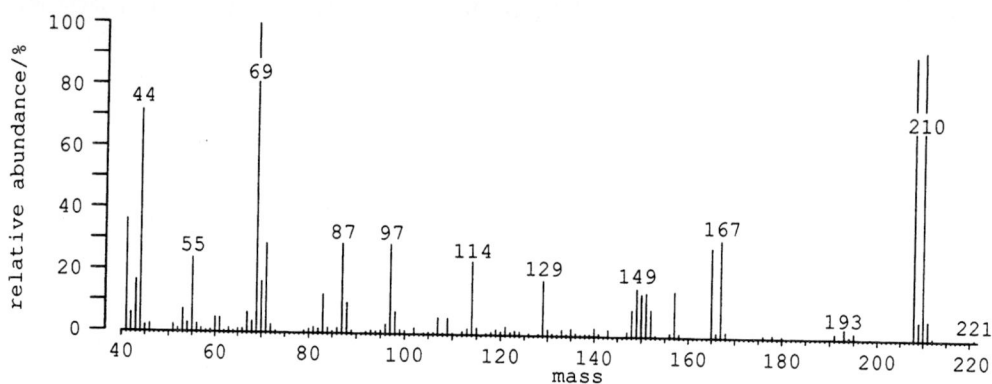

FIG 3   Carbrital

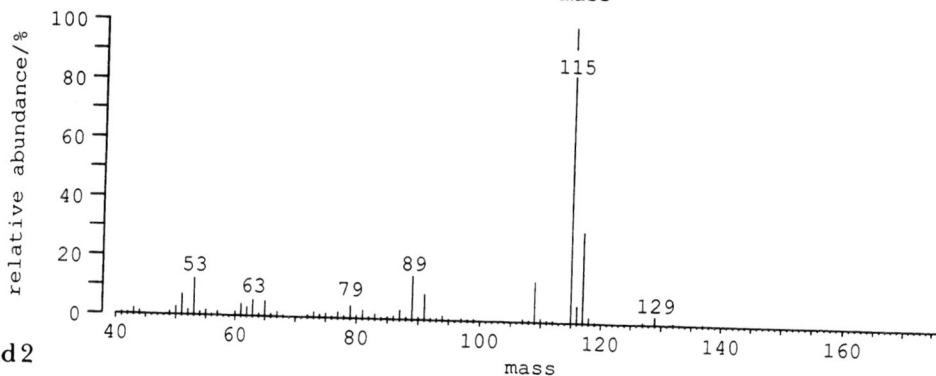

FIG 4   Compound 2

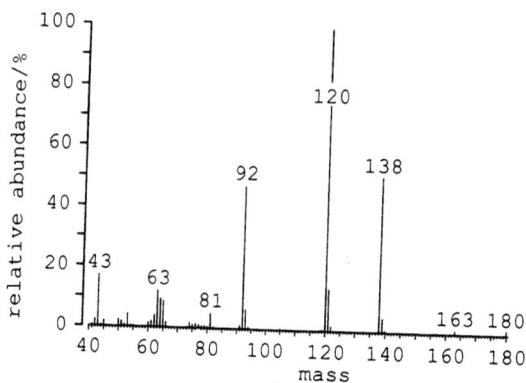

FIG 5   Aspirin

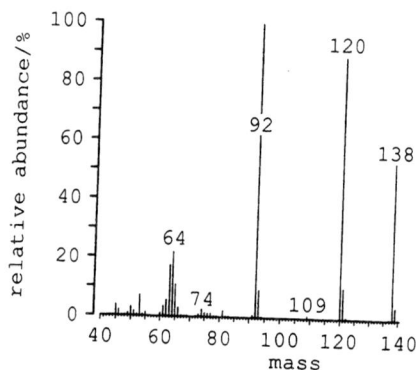

FIG 6   Compound 4

Questions

On low resolution mass spectra

For these questions use the data in table 2 (although if you compare the data for bromine, you will see that I have made some approximations).

TABLE 2

Isotope	^{12}C	^{13}C	^{19}F	^{35}Cl	^{37}Cl	^{79}Br	^{81}Br	^{127}I
Abundance/%	98.9	1.1	100	75	25	50	50	100

CONTINUED

3. The forensic chemist uses published tables to identify compounds from mass spectra. These commonly list the mass of each of the eight most abundant peaks, *in order of abundance*, the most abundant first. The listing for carbrital might be: 69 210 208 44 41 167 97 87. (In practice, the abundances are not perfectly reliable, especially below 10%. Another book's listing for carbrital is: 69 210 208 44 167 165 97 41. An expert would rely on the first four peaks, perhaps.)

(a) Construct an eight-peak listing for the second spectrum.

(b) Identify the compounds giving spectra 2 and 4 (figs 4 and 6), using table 3.

TABLE 3

Morphine	115	285	131	128	162	127	77	152
Placidyl	115	117	89	53	109	51	91	39
Prothoate	115	97	73	43	65	121	125	93
Tetrazolophthalazine	115	114	88	62	116	63	89	51
Benzocaine	92	120	165	65	137	0	0	0
Ethenzamide	92	120	105	148	150	121	133	65
Methylsalicylate	92	120	152	121	65	64	93	63
Salicylamide	92	120	137	65	121	39	64	53
Salicylic acid	92	120	138	64	39	63	121	65
Dicoumarol	92	121	120	65	162	63	93	64

Carbrital and compound 2 are both sedatives. Both are habit-forming, and neither is now prescribed in this country. In this case, it was assumed that the business man had obtained them abroad. There was a massive overdose present (as well as a large amount of aspirin) and a verdict of suicide was returned.

4. In another recent investigation, gas from the lungs of a teenager found dead on a canal towpath was found to contain CF_3Cl. (He had apparently been killed by inhaling aerosols. Although CF_3Cl is no longer used as an aerosol propellant, because it damages the ozone layer, aerosol sniffing still kills people.) The mass spectrum of CF_3Cl contains *two* M^+ peaks (table 4).

(a) Why are there two peaks of different masses?

(b) Explain their relative abundances.

(c) CF_3Br also has two M^+ peaks, but they are now of equal abundance. What are their masses and why are their abundances equal?

(d) CF_2Br_2 has three M^+ peaks (table 5).

Explain:

(i) why there are three M^+ peaks,

(ii) the abundance of these three peaks.

(e) CF_2Cl_2 has three M^+ peaks, of abundances 100, 67 and 11.

(i) What are the masses of the three M^+ peaks?

(ii) Explain their abundances.

(f) What would be the masses and abundances of the M^+ peaks arising from CF_2ClBr?

TABLE 4

Mass	104	106
Relative abundance/%	100	33.3

TABLE 5

Mass	208	210	212
Relative abundance/%	50	100	50

CONTINUED

5. Two of the compounds found in the stomach of our dead businessman contained one atom of a halogen: F, Cl, Br or I. Identify which compound contained which halogen and explain your decision. [Note that in the mass spectra of big molecules, the intensities may not be *exactly* as expected, but within ±5%.]

6. Aspirin's structure is shown in fig 7.

The peak at mass 180 in the mass spectrum of aspirin (fig 5) is M^+; that is, $C_9H_8O_4$ with one electron removed. The tiny peak at 45 is probably due to $[COOH]^+$. Suggest which positive ions are responsible for the peaks at:

   (a) 163 (note that the mass is 17 less than that of M^+),

   (b) 120,

   (c) 43,

   (d) 138 (*Hint*: Remove some atoms then put one back).

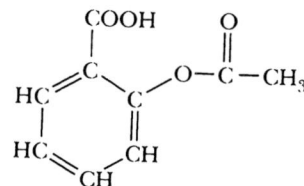

FIG 7

7. The last mass spectrum (fig 6) is that of an aspirin metabolite – a compound formed from aspirin in the body.

   (a) Suggest why the mass spectrum might lead us to expect the metabolite to be similar to aspirin itself.

   (b) Its formula is $C_7H_6O_3$. Suggest a structure for it.

   (c) Suggest what positive ion is responsible for the peak of mass 120.

   (d) The relative molar mass is 138. The peak of mass 139 arises because of the existence of ^{13}C. Explain (use table 2).

   (e) Why are there no obvious peaks of mass 140, 141 and so on?

   (f) The peak of mass 139 is about 8% of the size of the peak at 138. Explain why.

Sand castles and mud huts © 1991 Jeffrey Hancock, published by Hodder and Stoughton Educational

4. Quantum theory and cancer

If solid sodium chloride is put into a Bunsen flame, the flame goes yellow. Other compounds give different colours. For example, strontium chloride gives a red flame, barium chloride a green one. (This is used in the manufacture of fireworks.) Another way of producing coloured effects is by passing an electric current through a gas at low pressure; yellow street lamps use sodium vapour.

In each case, when energy is put into the substance (from the Bunsen or the electrical discharge), the electrons in the atoms are raised to higher energy levels. These electrons can then fall back to lower energy levels, releasing the extra energy again, mostly as light. (This light can be split up into an *emission spectrum*, fig 1, by passing it through a prism or diffraction grating.)

Or the reverse process can be made to happen, and an *absorption spectrum*, fig 2, produced. If white light is shone through sodium vapour, the sodium atoms absorb some of the light, and the electrons of the sodium atoms are raised to higher energy levels. When they fall back down again, the surplus energy is lost as heat. (For a detailed account of this, look up atomic spectra in any textbook.)

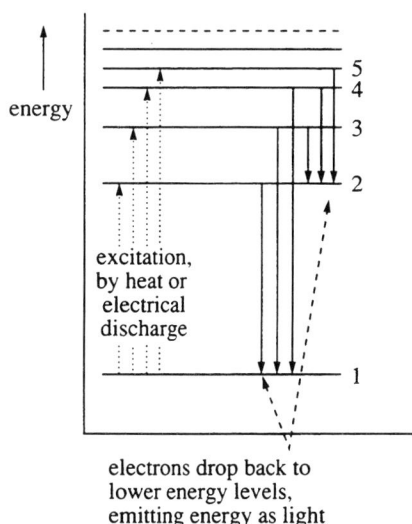

FIG 1   Emission spectra

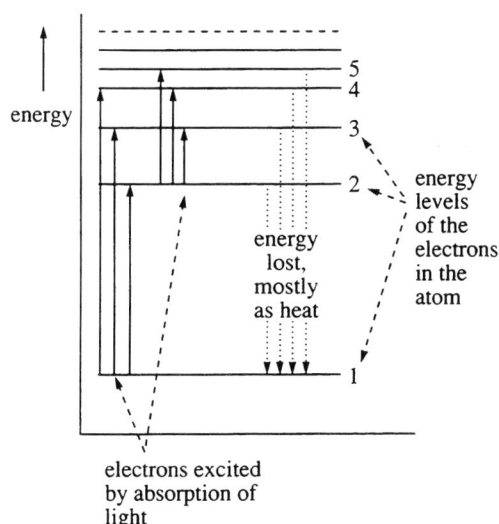

FIG 2   Absorption spectra

When sunlight hits the Earth's atmosphere, absorptions occur. The mechanism of these absorptions is similar to that for the atomic absorption spectra: as energy is absorbed, electrons are excited to higher energy levels. (The only difference is that the electrons are now in molecules and are undergoing transitions between *molecular* energy levels.) As a result, when the sunlight arrives at the Earth's surface most of the ultraviolet has been absorbed – 90% of its energy is in the visible and infra-red regions of the spectrum. But it is the ultraviolet (UV) that causes the most problems for biological systems. For convenience, it is divided into three regions: UVA 320–400 nm; UVB 290–320 nm and UVC 200–290 nm.

CONTINUED

Questions

1. Explain what is meant by:

   (a) 'infra-red' and 'ultraviolet',   (b) 'electrons are excited'.

2. Explain why:

   (a) different elements give rise to different flame colours,

   (b) sodium vapour absorbs yellow light of *exactly* the same wavelength as that emitted from a sodium flame.

3. Quantum theory suggests that the energy, E, associated with light of frequency, f, is given by $E = L \times h \times f$ kJ mol^{-1}; where L = Avogadro's number and h = Planck's constant (which has a value of 6.626×10^{-34} J Hz^{-1}).

   (a) Calculate:

   (i) the wavelength and hence the frequency of the middle of the UVA region of the spectrum (given that the wavelength, λ, and the frequency, f, are related by $c = f \times \lambda$, where c is the speed of light, 3×10^8 m s^{-1}),

   (ii) the energy of the midpoint of the UVA region.

   (b) Repeat for the midpoint of the UVB region.

   (c) Suggest why UVB reddens the skin faster than UVA.

   (d) There has been much concern expressed that destruction of the stratospheric ozone layer will let UVC reach the Earth's surface (see Section 18: Ozone problems, p. 67). Suggest why it is expected to be so damaging.

4. After excitation by light absorption, some compounds return to their ground state, losing the excess energy not as heat, but as light of a longer wavelength. This process – emission of light of a different wavelength – is called *fluorescence*, and is used in fabric brighteners added to washing powders, and in fluorescent paints. These absorb UV light and emit visible light.

   (a) Suggest why fluorescent compounds make the surface look brighter than usual.

   (b) Why isn't it possible to get a compound that absorbs energy in the visible region of the spectrum and emits it in the ultraviolet?

When skin is exposed to UV light, several things happen. The most obvious is erythema, or reddening, caused mainly by the UVB. At 40°N (approximately the latitude of New York or Rome) unprotected 'white' skin reddens in 15–20 minutes in summer. Longer exposure times cause burning. Other effects of UVA and UVB are less obvious. One is ageing: people who work outdoors or frequently expose themselves to the Sun develop a dry, leathery, wrinkled skin, caused by damage to the skin proteins. The skin cells' DNA can also be damaged by the absorption of UV light. It may be possible to repair this damage to the DNA or it may not, so there will be a range of possible consequences, from the death of the cell, or development of a cancer, to complete recovery. (A few people lack the enzyme that carries out this process; they are very susceptible to skin cancer.) There are other medical conditions caused by sunlight. For example, about a quarter of Caucasian (white-skinned) women suffer from polymorphic light eruption (PLE) within hours of exposure to the Sun. This is an unpleasant itchy rash which usually disappears if there is no further exposure. Recent research suggests that it is caused by UVA.

CONTINUED

For protection the skin synthesises a brown polymer called melanin which absorbs UVA, B and C and prevents them from penetrating the skin. (Albinos lack the enzyme that forms melanin and are forever vulnerable to the Sun.) This formation of melanin (tanning) occurs in two stages. First, colourless precursor molecules are made and stored in the skin. This is a very slow process, but then on exposure to UVA, the precursors are converted to melanin very rapidly. A tan can be produced within an hour. Further tanning requires the synthesis of more precursor molecules. This is accelerated by light in the UVB region, but is still very slow. In fact, it is far too slow to protect a white skin against unaccustomed amounts of Sun. So pale-skinned people who fly off to the Sun in summer, if they are wise will use some form of sunscreen until their tan has built up.

Questions

5. (a) Artificial sun beds emit UVA. They are often advertised as producing a suntan very fast. How do they work?

(b) Repeat treatments give a deeper tan – but only if there is a gap between treatments. If the treatments are repeated on consecutive days, there is no effect. Why?

6. (a) Polymorphic light eruption (PLE) is more common in summer. Suggest why.

(b) Sometimes further exposure to the Sun can cause the PLE to disappear. Suggest why.

(c) Some sunscreens may actually make PLE worse. Why?

Sunscreens contain compounds that absorb light in the crucial region of the spectrum. The absorbed energy is used to excite electrons from lower to higher molecular energy levels, and by careful choice of compound the absorption can be obtained in the desired position. There is a snag – most people want to come back from their holidays with a tan, but if they use too efficient a sunscreen, no UV light will get to the skin to produce it. Manufacturers quote a sun protection factor (SPF):

$$SPF = \frac{\text{time required for erythema with sun cream}}{\text{time required for erythema without sun cream}}$$

Your choice of sunscreen is linked to your skin type. If you have skin of type I (burns easily and never tans) you should use a screen with a high SPF, while for a skin of type V (rarely burns, tans deeply) a lower SPF is adequate. Unfortunately, the SPF is at present measured by erythema, and erythema is produced by UVB. The SPF therefore gives no guidance for UVA protection. Many sunscreens give little or no protection against UVA. Doctors now suggest that UVA protection is important.

Questions

7. Two sunscreen creams were tested on three volunteers. Table 1 on the next page gives the times taken for erythema to develop without and with the two sunscreens.

(a) Calculate the mean SPF values for each sunscreen. Suggest a reason why there is a range of values.

(b) Which of the two sunscreens:

(i) would you recommend for a fair skinned person?

(ii) would result in faster tanning?

CONTINUED

TABLE 1

Volunteer	Time for erythema to develop, with:		
	no cream /minutes	cream 1 /hours	cream 2 /hours
A	12	3.25	2.25
B	20	5.67	3.25
C	16	4.50	3.00

8. The UV spectra of several sunscreen constituents are given in figs 3–6. The y-axis gives the amount of light absorbed; the x-axis, the wavelength in nanometres.

(a) Suggest three characteristics that a chemical for use in a sunscreen should possess. (In addition to absorbing UV light!)

(b) Point out three features that the structures of the four molecules have in common.

(c) Which of the compounds given would protect the skin against UVA, and which against UVB?

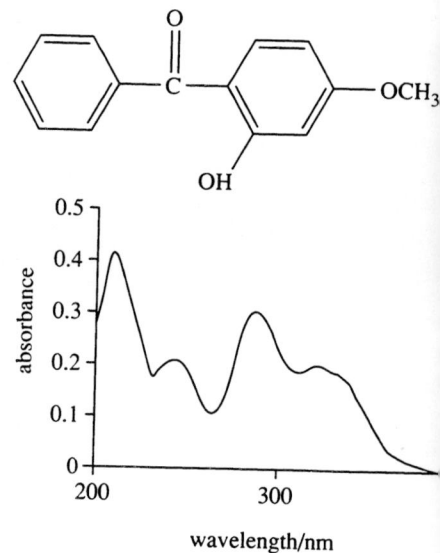

FIG 3   Benzophenone-3

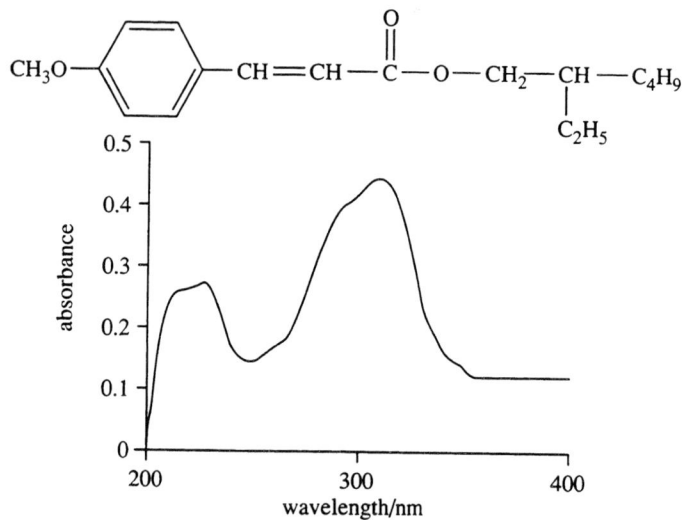

FIG 4   Octyl methoxycinnamate

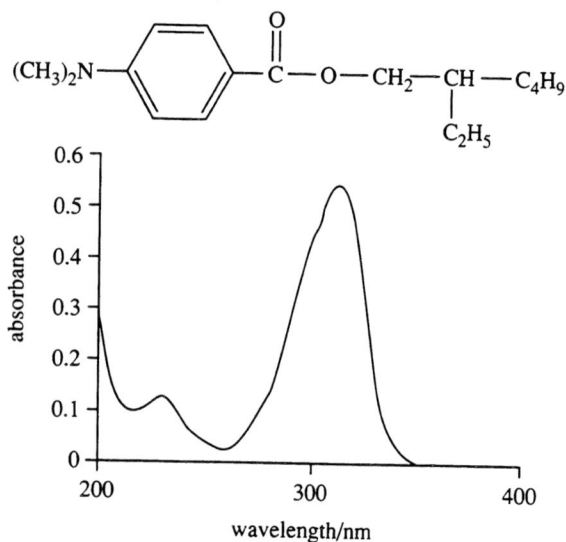

FIG 5   Octyldimethyl paba

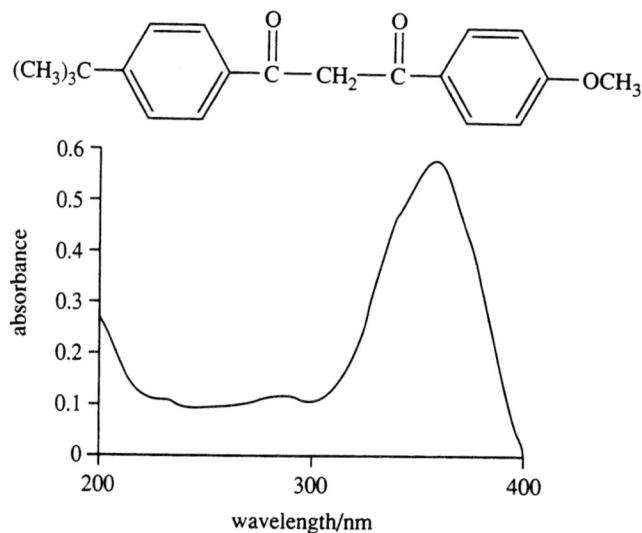

FIG 6   Butyl methoxy dibenzoyl methane

CONTINUED

Sand castles and mud huts © 1991 Jeffrey Hancock, published by Hodder and Stoughton Educational

9. The amount of light absorbed at a particular wavelength, the *absorbance*, is given by the Beer-Lambert law:

$$\text{absorbance} = \varepsilon \times c \times l$$

where c = concentration in $mol\,dm^{-3}$,
  l = the path length of the light (that is the thickness of the solution that the light has to travel through) in cm,
  ε = a constant for the particular compound.

(a) Octyl methoxycinnamate absorbs light best at 311 nm. Its absorption spectrum is given in fig 4. Read off a value for the absorbance at 311 nm.

(b) Use it to calculate a value for the constant ε for this compound. (The concentration of the solution used to get the spectrum was $5.24 \times 10^{-3}\,g\,dm^{-3}$, the relative molar mass of octyl methoxycinnamate is 290 and the path length was 1 cm.)

(c) I have a sunscreen which contains 6% octyl methoxycinnamate (that is, $60\,g\,dm^{-3}$). I spread this on my skin, say to an average thickness of 0.01 mm. Use your value for ε to calculate the absorbance of this sunscreen.

(d) But how much light gets through it to my skin? The absorbance is related to the intensity of light actually getting through by:

$$\text{Absorbance} = \log_{10}[I_0/I]$$

where I_0 = intensity of the light incident on the substance, and
  I = the intensity of the light actually transmitted.

Use the value for the absorbance that you obtained in (c) to estimate what fraction of the sunlight hits the flesh (that is, calculate a value for I/I_0).

5. Bitter-sweet

(Do not taste any chemical in the laboratory)

We do *most* of our tasting with our tongue, on which there are some 9000 taste buds. Each taste bud consists of many special cells, connected to nerve ends. These taste cells disintegrate after about a week, and are replaced. Inevitably this replacement of dead cells by new ones gets less efficient as we get older, so that our sense of taste declines.

There are thought to be four basic flavours, detected on different regions of the tongue (fig 1).

(There's more to it than that, though. Smell and chewing also play a part. For example, if you hold your nose and don't chew, it is difficult to tell an apple from an onion!)

But how do the taste buds work? It seems that they contain specific receptor molecules, probably proteins. These molecules contain regions of a particular shape, into which small 'tasty' molecules exactly fit. Just as the correct piece will fit into a jigsaw puzzle, so a sweet molecule will fit the three-dimensional receptor site on the protein in the 'sweet' area of the tongue, but a tasteless one will not. This mechanism probably causes some change in the protein, so triggering a nerve impulse to the brain.

So what *sort* of molecules fit into this receptor site?

It seems that a sweet molecule must contain a very specific arrangement of atoms (fig 2). It must have two *very electronegative* atoms (A and B in fig 2). These are usually nitrogen or oxygen and they probably form hydrogen bonds with the receptor molecule. These hydrogen bonds can form only if the B atom and the hydrogen on the A atom are 0.30 nm apart (that is, in fig 2 the distance marked 'z' must be 0.30 nm). When they are this distance apart, the sweet molecule and the protein can interact, a nerve is stimulated and we detect the sweetness.

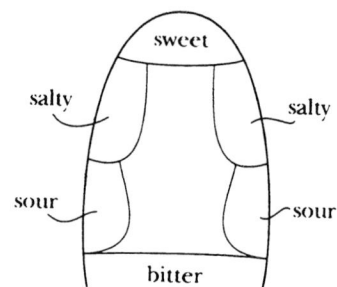

FIG 1

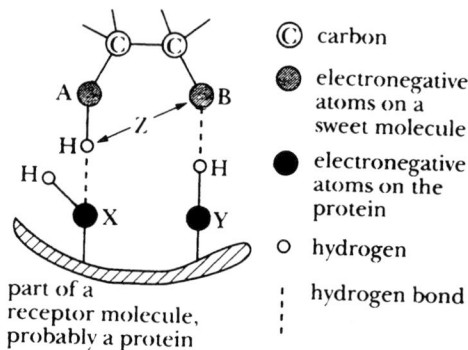

part of a receptor molecule, probably a protein

FIG 2

Questions

On the relationship between flavour and molecular shape

1. What is meant by:

   (a) 'specific receptor molecules'?

   (b) 'electronegativity'?

2. (a) Draw a diagram to show all the bonding electrons in methane, CH_4.

   (b) The $H-C-H$ bond angle in methane is 109°28′. The $H-N-H$ angle in ammonia is about 107°. Explain the difference.

   (c) Draw a diagram of a water molecule and label the $H-O-H$ bond angle.

3. (a) Ethane-1,2-diol (CH_2OH-CH_2OH) is probably the simplest substance that tastes sweet. What values would you expect for the bond angles labelled α, β and γ in fig 3? (Be careful – think about the shapes of the methane and water molecules.)

   (b) Use your data books to find values for the lengths of the $C-H$, $C-C$, $C-O$ and $O-H$ bonds.

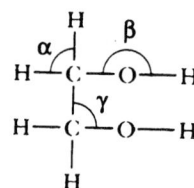

FIG 3

CONTINUED

Sand castles and mud huts © 1991 Jeffrey Hancock, published by Hodder and Stoughton Educational

(c) Use a ruler and protractor to make a scale drawing of the important part of the molecule (as far as sweetness is concerned), using a scale of $5 \times 10^8 : 1$, that is 50 cm = 1 nm. (Make sure you draw the angles correctly!)

(d) You should realise that the critical distance for sweetness, the one marked 'z' in fig 2, can take a range of values in ethane-1,2-diol. Why is this? What are the maximum and minimum values z can take? How can it become exactly 0.30 nm, as required for the molecule to taste sweet? (It may help if you use a piece of wire to make a scale model of the important part of the molecule. Remember that there is free rotation about all single bonds.)

(e) Sketch a molecule of ethane-1,2-diol interacting with the receptor protein.

4. (a) What is a hydrogen bond?

   (b) If a hydrogen bond is to form, of the type: A−H . . . X (fig 2):

      (i) atom A must be of high electronegativity. Why?

      (ii) atom X must be of high electronegativity. Why?

5. Suggest explanations for the following (fig 4).

   (a) Ethane, $CH_3−CH_3$, is not sweet.

   (b) Ethanol, $CH_3−CH_2OH$ is not sweet.

   (c) 2-aminoethanol is sweet.

   (d) Propane-1,3-diol is not sweet.

   (e) 2-chloroethanol is sweet, but not as sweet as ethane-1,2-diol.

   (f) 2-hydroxyethanethiol is not sweet.

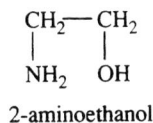

2-aminoethanol

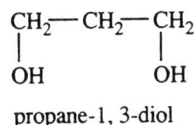

propane-1, 3-diol

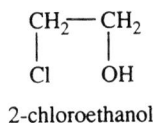

2-chloroethanol

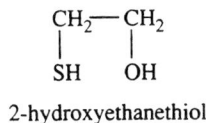

2-hydroxyethanethiol

FIG 4

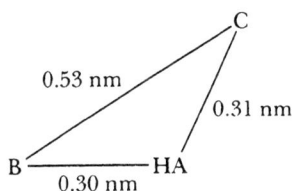

FIG 5

But this is only part of the story. Very sweet molecules also have a *non-polar* region, C, arranged an exact distance from the AH and the B groups (fig 5).

CONTINUED

This 'C' region presumably interacts with a non-polar receptor site on the protein. These three regions can be identified on most sweet molecules. For example, glucose is shown in fig 6 (not all the H atoms are shown).

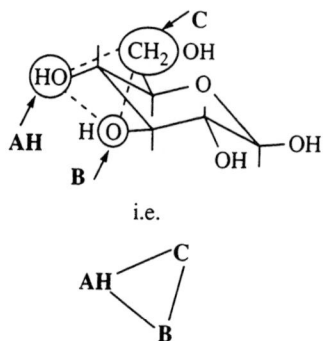

i.e.

FIG 6

A similar theory has been put forward to explain the bitter taste of some substances, except that the distance between the AH and B groups now has to be about 0.15 nm.

Questions

6. In fig 7 the structures of various synthetic sweeteners are given. Copy the structures and indicate on your diagram the AH and B groups, and the non-polar C region.

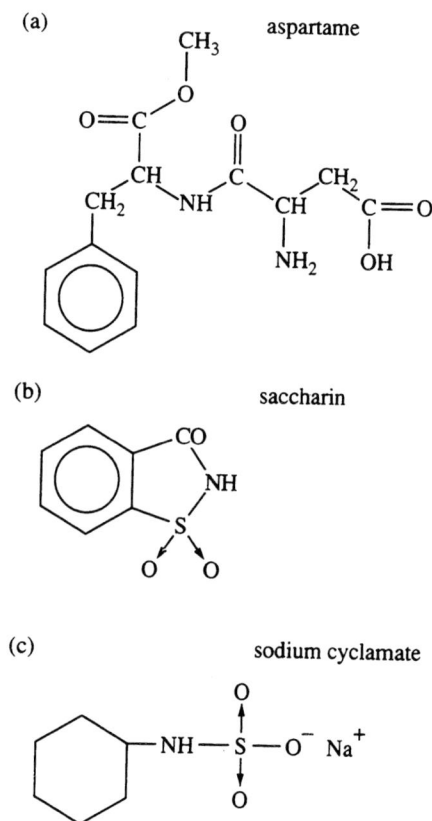

(a) aspartame

(b) saccharin

(c) sodium cyclamate

FIG 7

CONTINUED

Sand castles and mud huts © 1991 Jeffrey Hancock, published by Hodder and Stoughton Educational

7. 'Bitrex' (fig 8) is apparently the bitterest substance known. It is added to lavatory cleaners to encourage people not to drink them. Which part of the 'Bitrex' molecule is responsible for its bitter taste?

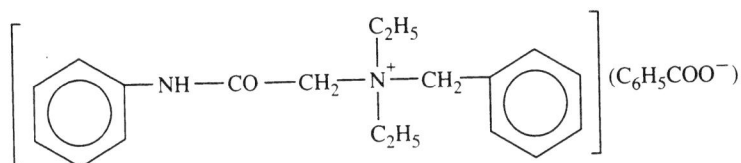

FIG 8

8. Starch is a polymer, consisting of thousands of glucose units joined together. Why does it not taste sweet?

9. Cyclopentane-1,2-diol's structure is shown in fig 9.

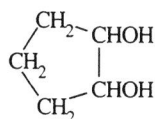

FIG 9

(a) What values would you expect for the angles in the ring (which can be taken to be a regular pentagon)?

(b) What bond angles do you expect to find around a saturated carbon atom?

(c) Why is the molecule planar?

(d) Can you suggest why it does not taste sweet at all? (Refer back to your answer to question 3.)

6. Anaesthetics

A hundred and fifty years ago, surgical procedures were pretty harrowing. Operations were quick and nasty.

Syme's record for a mid-thigh amputation was alleged to be nine seconds, including the patient's left testicle and the forefinger of the chief assistant.

In 1846, William Morton, a Massachusetts dentist, administered ethoxyethane vapour to a patient so that the surgeon could remove a tumour from his neck. This anaesthetic was rapidly adopted and dinitrogen oxide and trichloromethane were soon in use as well.

A number of other anaesthetic gases were discovered during the next hundred years. But by about 1940, their disadvantages were becoming clear. Dinitrogen oxide is not very potent, trichloromethane causes liver damage and ethoxyethane is highly flammable. (This became an acute problem with the increasing use of electrical equipment in the operating theatre. Several deaths were caused by explosions during operations.)

In 1950, Imperial Chemical Industries set up a programme to find a better anaesthetic. A company embarking on expensive research like this will not just make new substances and test them; there are far too many possible compounds. Instead, the research workers use the best available theory to try to predict what sort of compounds are likely to be useful.

So how does an anaesthetic work? The gas is inhaled into the lungs, where it dissolves in the blood. This carries it to the brain, where it acts to suppress pain. We might represent it like this:

anaesthetic gas $\rightleftharpoons$ lungs $\rightleftharpoons$ blood $\rightleftharpoons$ brain $\rightleftharpoons$ pain suppression [1]

Questions

On intermolecular forces

1. Explain what is meant by:

    (a) 'Dinitrogen oxide is not very potent . . .',

    (b) '$\rightleftharpoons$'.

2. If an anaesthetic is to be transported from the lungs to the brain, it presumably must be soluble in the blood (that is, in water). Substances dissolving well in water are either ionic or polar.

    (a) What is meant by 'ionic'?

    (b) Unfortunately, ionic substances are usually solids at body temperature (so they can't easily be inhaled). Why are they solids?

    (c) Dinitrogen oxide (N_2O), trichloromethane ($CHCl_3$) and ethoxyethane ($C_2H_5OC_2H_5$) are all polar. Explain what is meant by the word 'polar', and why each of these compounds is polar. (None of the molecules is linear.)

3. If an anaesthetic has a high solubility in water (or blood), it acts very rapidly, and lasts a long time. In other words, it is eliminated from the body slowly. Explain why:

    (a) it acts fast,   (b) it is long lasting.

CONTINUED

Sand castles and mud huts © 1991 Jeffrey Hancock, published by Hodder and Stoughton Educational

But *does* the effectiveness of an anaesthetic depend on its solubility in water? The effectiveness of an anaesthetic gas is usually measured by the pressure of the gas needed to get it to work. A low pressure is required for a good anaesthetic, while a high pressure is needed to get a poor anaesthetic to work. We want to know if there is a relationship between the pressure an anaesthetic requires, and its solubility in blood or water. Is the pressure required for anaesthesia proportional to the solubility of the gas in water, perhaps raised to some power? If so:

$$\text{pressure of anaesthetic gas} = \text{constant} \times (\text{solubility in water})^n \quad [2]$$

that is,
$$P = k(\text{water solubility})^n \quad [3]$$

where k and n are constants. Now if we take logarithms throughout the equation:

$$\log P = \log k + n . \log (\text{water solubility}) \quad [4]$$

This is of the same form as:

$$y = c + m . x$$

which is the equation of a straight line graph. So if the effectiveness of an anaesthetic *is* related to its solubility in water, a graph of log (pressure of anaesthetic) plotted against log (water solubility) should be a straight line (of gradient n) (fig 1).

Two things are clear. There is a relationship. The higher the anaesthetic's solubility in water, the lower the pressure needed for anaesthesia. But the correlation is pretty poor – the points aren't very close to the straight line.

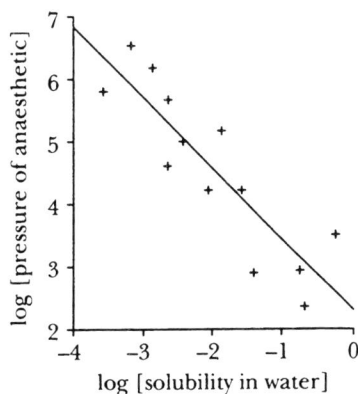

FIG 1

Questions

4. The anaesthetic pressures for various gases are given in table 1.

TABLE 1

Gas	N_2	Ar	Kr	Xe
Anaesthetic pressure/kPa	2940	2027	294	86

(a) Which gas is the most effective anaesthetic?

(b) Suggest what sort of values you would expect for the anaesthetic pressures of neon and helium.

(c) When working at great depths, divers can experience the problem of nitrogen narcosis (anaesthesia). This should not be confused with decompression sickness (the bends), which is caused by the rapid release of gas dissolved in the blood, with the consequent formation of bubbles. Suggest why nitrogen narcosis occurs only at great depths.

(d) Why can it be prevented by breathing a helium-oxygen mixture instead of air?

5. Fig 1 shows that the effectiveness of an anaesthetic depends on its solubility in water.

(a) The more polar a gas molecule is, the more soluble it is in water. Explain this in terms of intermolecular forces.

(b) CCl_2F_2 is more soluble in water than CF_4. Why is this? [Remember that both of them are tetrahedral molecules.]

(c) So if water solubility was the only important factor, which would be the better anaesthetic, CCl_2F_2 or CF_4?

CONTINUED

(d) In the same way, suggest which of the following would be more water soluble, and hence the better anaesthetic:

    (i) $CHCl_3$ or $CHClF_2$?

    (ii) CF_3-CH_3 or $CF_3-CHClBr$?

Once the anaesthetic has dissolved in the blood, it is transported throughout the body. Then it has to get into the brain where it can act. To cross the brain membranes, it will probably have to dissolve in the lipids, which are the major constituents of cell membranes. Lipids are non-polar organic compounds, such as fats or oils. In experimental work, olive oil is often taken to be a convenient and typical lipid. Does the effectiveness of an anaesthetic depend on its solubility in lipids? Fig 2 is a plot of the logarithm of the anaesthetic pressure against the logarithm of the anaesthetic's solubility in oil, for 22 different anaesthetics, in three different animals.

The correlation is excellent. This is a good straight line graph. So it seems that a major factor governing the effectiveness of an anaesthetic is its solubility in the lipid molecules of cell membranes.

What does the anaesthetic *do* once it has reached the brain? Must it have a particular shape, polarity, or size to interact with a particular receptor site in the brain? There is a huge range of anaesthetics, from simple molecules, like nitrogen or argon, to quite complex molecules such as ethoxyethane. Can we possibly imagine a specific process that will work for all of these? It seems unlikely. In fact, nobody yet knows exactly how anaesthetics work, but it seems that their effect on cell membranes is important.

Armed with these ideas, ICI formulated some guidelines for the new anaesthetic. The compound had to be somewhat soluble in water, and soluble in olive oil. It also had to be chemically stable, so it wouldn't break down before use. Organic compounds containing fluorine are very stable, so they were the first choice, and since the fully fluorinated hydrocarbons do not burn, that overcame the flammability problem too.

Some twenty or so new fluorine-containing compounds were made and tested on mice, rabbits and finally dogs. Some of them caused convulsions, some had adverse effects on the animal's blood pressure and so on. Just one compound from this first batch was any good – but it was very good. Halothane, $CF_3-CHClBr$, had its first human trial on 20 January 1956 and was in commercial production eighteen months later. It is now the most widely used anaesthetic in the world.

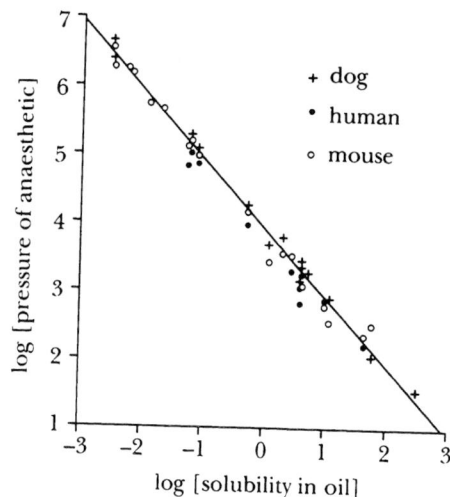

FIG 2

Questions

6. Fig 2 shows that the effectiveness of an anaesthetic depends, to a large extent, upon its solubility in oils or lipids.

    (a) If a compound is soluble in oils or lipids, what sort of bonding will it have, and what can you say about the forces between the molecules?

    (b) Explain how these intermolecular forces arise. (Use a simple anaesthetic like xenon as an example.)

    (c) Olive oil is a convenient substance that is similar to lipids. Can you suggest two other compounds that could be used in this way?

7. (a) CF_4 is more soluble in olive oil than CCl_2F_2. Why?

    (b) If solubility in oil was the only important factor, which of these two compounds would be the more effective anaesthetic?

CONTINUED

Sand castles and mud huts © 1991 *Jeffrey Hancock, published by Hodder and Stoughton Educational*

(c) Similarly, which would be more soluble in oil, and hence the better anaesthetic:

(i) $CHCl_3$ or $CHClF_2$?

(ii) CF_3-CH_3 or $CF_3-CHClBr$?

(d) In fact, the anaesthetic pressure for CF_4 is 7.3 kPa, and for CCl_2F_2 it is 41 kPa. Which is the better anaesthetic? Compare your answers to questions 5(c) and 7(b). Which factor, water solubility or oil solubility, is more important in an anaesthetic?

(e) Compare your answers to questions 5(d) and 6(c), and predict which would be the more effective anaesthetic:

(i) $CHCl_3$ or $CHClF_2$?

(ii) CF_3-CH_3 or $CF_3-CHClBr$?

8. The ICI scientists believed that if a compound was very unreactive – like fluorinated hydrocarbons – it was not likely to be toxic.

(a) Why was this a reasonable assumption?

(b) Find one piece of evidence from the passage which suggests that it was not completely true.

9. Determine an approximate value for the gradient of the graph in fig 2, and hence determine a value for n in the equations 2–4. Comment on your result.

7. *Sand castles and mud huts*

Pack wet sand into a small bucket, press it down firmly, then invert the bucket on the top of the castle. Give the bucket a little shake and lift it gently. You have a turret to support a flagpole. Until the tide comes in, of course, or until the sand dries out. *Dry* sand is useless as a building material. It turns to powder and slides down into a heap, unlike clay or mud which as they dry get harder. Millions of people live in houses made of dried mud.

Why is there this difference between sand and clay?

Let us simplify a little, and say that sand is silica (silicon dioxide, SiO_2). This is a good simplification. Although there will be impurities in the sand – which turn it to the usual golden colour – they make no difference to the theory. The SiO_2 lattice is an extended covalent lattice, with each silicon atom covalently bonded to four oxygen atoms, and each oxygen to two silicons (fig 1).

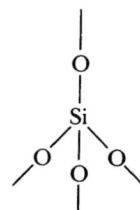

FIG 1

What happens at the edge of the crystal? Perhaps some of the silicon atoms are only bonding to three oxygens; in other words, they have empty spaces in their orbitals. When the crystals get wet, water molecules may use these vacancies to bond to the surface of the silica crystals (fig 2).

These water molecules can then hydrogen bond to other water molecules (fig 3) thus holding the grains of sand together, and enabling you to build your castle. When the water evaporates, there is nothing left to hold the sand crystals together and they fall apart.

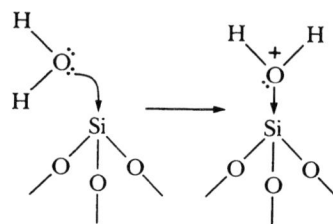

FIG 2

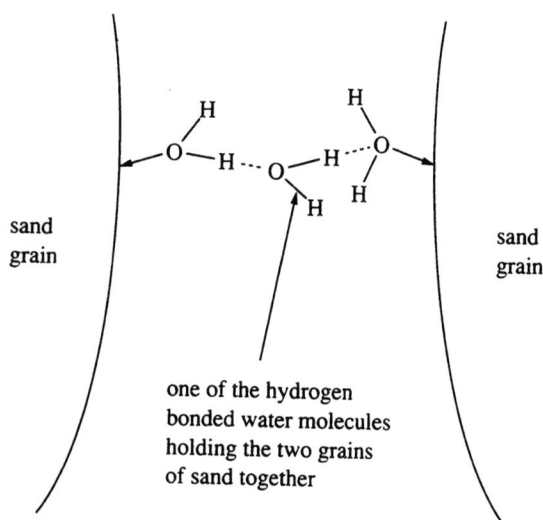

sand grain

sand grain

one of the hydrogen bonded water molecules holding the two grains of sand together

FIG 3

To make a good sand castle, the sand must be wet – but not too wet, or the sand just settles to the bottom of the bucket. Because the sand grains are very irregular in shape, the hydrogen bonding must be irregular, and so it isn't strong enough to bind much water in the spaces between the grains of sand. Indeed, the sand particles will be virtually touching. They will be close-packed, like marbles in a bag, or the atoms in a metal lattice.

CONTINUED

Questions

On atomic structure and intermolecular forces

1. (a) Give the electronic configurations of silicon and oxygen.

   (b) Explain how silicon and oxygen bond.

   (c) Why does silicon bond to four oxygen atoms, but oxygen to only two silicon atoms?

   (d) What is meant by 'an *extended* covalent lattice'?

   (e) Explain why the silicon atoms 'have empty spaces in their orbitals'.

2. Explain:

   (a) why water molecules are polar, and

   (b) how they hydrogen bond to each other, for example, in ice.

3. (a) What does the symbol '→' mean in fig 2?

   (b) How many electrons does the silicon atom have before and after bonding to the water molecule? How is this possible?

4. (a) If pure silicon dioxide is added to pure water, the pH of the water falls. Can you explain this, starting from fig 2?

   (b) If the particles of silica are quite big, the pH of the water falls very little. But if the silica particles are very small, the pH can drop below 5. Why?

5. Suggest why:

   (a) a wet beach is firm to walk on,

   (b) a sand castle doesn't shrink as it dries.

Clays are chemically more complicated, and there is a vast range of different ones. Kaolinite, $Al_4(OH)_8Si_4O_{10}$, is one of the simplest. Just as with silica, there are incompletely bonded atoms at the surface of the clay particle. When the clay gets wet, the water molecules will interact with the clay surface just as they did with the sand grains. There are two differences, however.

First, the clay particles are much smaller: with a length of, say, 10^{-8} to 10^{-7} m, compared to 10^{-3} or 10^{-4} m for sand. This means that the clay particles have a much larger surface area. Many more atoms are therefore on the surface of the clay crystals, so there are *many more water molecules attached*.

Second, whereas the sand particles are an irregular spherical shape, the clay particles are more regular, and flat, like plates (fig 4). So the hydrogen bonding can be more regular than in wet sand. The water molecules are therefore held more strongly, and so more of them can be held between the clay particles than between sand particles. Because there is more water, the particles in clay can be loosely packed. So wet clay can be deformed easily by pressure, because there is so much water that the clay particles can slip over one another.

Wet clay shrinks as it dries, until it reaches a certain water content, when the volume stays roughly constant. This is presumably because the wet clay holds so much water that the particles aren't touching, and shrinkage occurs until they do. In order to move the particles in dried clay you have to break many hydrogen bonds. And because these hydrogen bonds are regular you have to break them *all at once*. This will

CONTINUED

clearly be much harder to do than to break the fewer hydrogen bonds in sand, one by one. (It's rather like undoing a zip fastener by pulling it apart, compared to opening it from the end.)

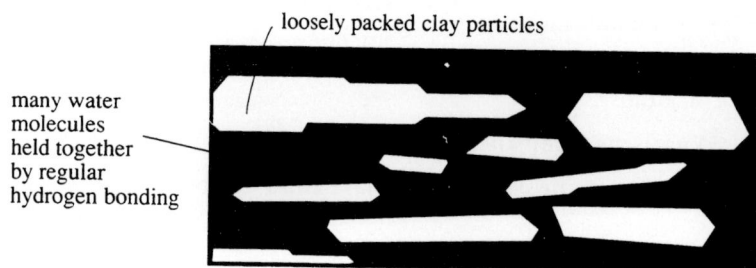

loosely packed clay particles

many water molecules held together by regular hydrogen bonding

FIG 4

Questions

6. (a) If you walk across wet clay, your feet sink in. A potter can shape the wet clay by pressing it with his hands. How are these things possible?

   (b) Wet sheets of glass, such as microscope slides, stick together very strongly, but they are quite easy to *slide* apart. Why?

7. Fig 5 shows the volume of a sample of kaolinite as it was completely dried in an oven.

   (a) Why does the volume fall?

   (b) If the clay is left to dry in air, rather than in an oven, the water loss *stops* when the water content has fallen to about 16%. Why do you suppose this is so?

   (c) What is the significance of this 16% of water?

   (d) Why is dry clay hard and strong? In other words why can clay particles stick together in a way that sand particles can't?

FIG 5

8. Some clays, for example bentonite, form *thixotropic* mixtures with water. A bentonite-water mixture (containing only 0.05% bentonite) forms a jelly-like substance which won't pour out of a test tube even if you turn it upside down. If you give the tube a sharp tap, the mixture becomes completely liquid. On standing it becomes solid again.

   (a) Suggest how thixotropy arises. Why will a bentonite-water mixture form a jelly, turn to a liquid if tapped sharply, then reform the jelly on standing?

   (b) Some paints ('non-drip') are thixotropic. Why is this property useful?

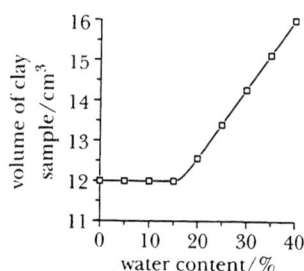

8.   Moles and things

Use the relative atomic masses from your periodic table or data book. Some other useful bits of data:

$$1 \text{ tonne} = 10^3 \text{ kg} = 10^6 \text{ g}$$

$$1 \text{ m}^3 = 10^3 \text{ dm}^3 = 10^6 \text{ cm}^3$$

$$\text{Avogadro's number} = 6.0 \times 10^{23}$$

1 mole of any gas occupies about 24 dm^3 at 298K and 10^5 Pa.

Questions

1. It is thought that the universe came into existence between 10 and 20 billion years ago (American billions, that is, between 1 and 2×10^{10} years ago). How long ago was the universe a mole of seconds old? Make a quick guess/estimate, then calculate it more carefully. (Ignore leap years!)

2. The Apollo moon mission landed two men on the moon in 1969. The rocket that controlled the descent and took them off again was fuelled by a mixture of two liquids: methylhydrazine (CH_3NHNH_2) and dinitrogen tetraoxide (N_2O_4) that ignited on contact.

(a) Devise an equation for the reaction. The products are nitrogen, water and carbon dioxide.

(b) The descent to the lunar surface required about 4.5 tonnes of dinitrogen tetraoxide, and taking off again about 1.5 tonnes. What was the total mass of methylhydrazine needed?

3. The solid booster rockets of the space shuttle are fuelled by a mixture of aluminium and ammonium chlorate (VII) (ammonium perchlorate, NH_4ClO_4).

(a) If no other reagents are involved, and the products are nitrogen, water, hydrogen chloride and aluminium oxide, devise an equation for this reaction.

(b) Each launch consumes about 160 tonnes of aluminium. What mass of ammonium chlorate (VII) must be carried? What mass of HCl gas is produced in the atmosphere above the Cape Canaveral launch pad?

4. Estimate the volume of ethene, measured at room temperature and pressure, that is needed to make a polythene washing-up bowl.

5. A ten pence coin is made of cupro-nickel: 75% copper, 25% nickel.

(a) Estimate the number of atoms in it, assuming:

(i) that the mass of a coin is 11.25 g, *or*

(ii) that the coin has a diameter of 2.75 cm and a thickness of about 0.25 cm, and that the radius of a copper or nickel atom is about 0.128 nm. (The volume of a sphere of radius r is given by $V = 4/3\pi r^3$; that of a cylinder of radius r, height h, by $V = \pi r^2 h$.)

(b) Which estimate do you think is the most reliable? Explain why.

CONTINUED

6. An electric hand drier blows out air heated by a 2.40 kW heater. After 30 seconds it turns itself off.

(a) Why does the air blast feel cold for some time, before getting hotter?

(b) Estimate the mass of water that the machine can evaporate off your hands in 30 seconds. (Remember that $1\,W = 1\,J\,s^{-1}$. The enthalpy of vaporisation (or latent heat of evaporation) of water is $44.0\,kJ\,mol^{-1}$.)

7. A marathon runner requires about 280 kJ for every kilometre of a 42 km race.

(a) If he gets the energy by burning glucose, calculate the mass of glucose required for the race. (Glucose, $C_6H_{12}O_6$, releases 2800 kJ per mole on combustion.)

(b) This heat has to be dissipated somehow. If evaporation of sweat was the only mechanism (it isn't, of course), what mass of water would have to be evaporated during the race? (The enthalpy of vaporisation (or latent heat of evaporation) of water is $44.0\,kJ\,mol^{-1}$.)

8. Carbon dioxide is likely to become an increasingly serious atmospheric pollutant, contributing to the greenhouse effect, and a rise in the Earth's temperature.

(a) Estimate the amount of CO_2 produced by a runner during a marathon. (The marathon is about 42 km long. Assume that exhaled air contains 4% CO_2, that he breathes once every two seconds and that the usable volume of the athlete's lung is $4\,dm^3$. A good time for a marathon is about 2½ hours.)

(b) What about a car? How much CO_2 does it produce in the same distance (which is close to the daily average for a family car)? At a constant speed of 90 kph (56 mph), the Peugeot 205 XL uses 4.5 litres of petrol per 100 km. Assume for convenience that petrol is octane (C_8H_{18}, density = $0.703\,g\,cm^{-3}$).

(c) Are diesel engines an answer to the problem of pollution? How much CO_2 does a diesel engine produce? At a constant 90 kph, the Peugeot 205 XLD uses 3.9 litres of fuel per 100 km. Assume that diesel fuel is octadecane ($C_{18}H_{38}$, density = $0.777\,g\,cm^{-3}$).

9. A newspaper reported last year that a cow produces 250 g of methane a day.

(a) Is this another journalistic exaggeration? What would the volume of the gas be?

(b) Methane is another 'greenhouse gas', and is about 30 times more effective at trapping heat than CO_2. Your answers to 8(b) and (c) will have given you a rough idea of a car's daily CO_2 output. How does a cow compare? (Neglect the CO_2 that the cow breathes out.)

10. The female silkworm moth attracts its partner by emitting chemicals (pheromones) onto the breeze. A female emits about $2\,\mu g$ (that is, $2 \times 10^{-6}g$) of a substance with a relative molecular mass of about 150. Some time later, a male arrives, having detected her 'smell' about 1.5 km away. Estimate how many molecules of the pheromones he *originally* detected. (Assume that the whole $2\,\mu g$ was spread evenly throughout a hemisphere of radius 1.5 km, and that the male moth sampled $1\,cm^3$ of air at a time. The volume of a sphere, of radius r, is given by $V = 4/3\pi r^3$.)

Sand castles and mud huts © 1991 Jeffrey Hancock, published by Hodder and Stoughton Educational

9. Import–export

This isn't a new problem. In 1661, John Evelyn complained to Charles II about London's air. To no avail – it took the great smog of December 1952 (which killed 4000 people in five days) to get the Clean Air Act passed in 1956. This act restricted only smoke emissions. There is still little control on gases.

A major pollutant is sulphur dioxide (SO_2). This is a colourless gas which smells slightly sweet at low concentrations, but is vilely choking at higher ones. Sulphur emissions result not only from human activities; natural processes also put large quantities into the atmosphere. In 1977 it was estimated that 500 million tonnes of sulphur entered the atmosphere annually, of which industry provided about 200 million tonnes. Most of this (60% of the total) was SO_2 released by burning coal and coke. Although using tall chimneys helps to avoid high concentrations of SO_2 in the area around a factory, it merely disperses the problem. In 1983, of the 225 000 tonnes of sulphur which fell on Norway, 92% of it was imported. (In contrast, 80% of the 847 000 tonnes that dropped on Britain was home-made.)

What happens to all this airborne sulphur?

Much of it is breathed in by you and me. The TLV (Threshold Limiting Value, the level considered safe to breathe for eight hours) is five parts per million (ppm), and even ten ppm seems to cause no ill effects in most people. But when SO_2 and smoke are inhaled together – as happens in most air pollution incidents – some people experience problems at concentrations as low as *0.75 ppm*. Plants are even more sensitive. Table 1 contains data for a forest in New York State.

TABLE 1

Site	Position in relation to source of SO_2	Percentage of 1-year-old foliage damaged	Percentage of 2-year-old foliage lacking	SO_2 concentration /ppm
1	*19 miles NE*	*77.9*	*20.6*	*0.045*
2	*25 miles NE*	*55.6*	*15.2*	*0.017*
3	*40 miles NE*	*16.7*	*9.1*	*0.008*
4	*93 miles W*	*2.1*	*3.9*	*0.001*

The maximum permitted value for the *average* annual SO_2 concentration in air is 0.028 ppm or 0.042 ppm (depending on the smoke levels). This value was laid down by the European Community and will be in operation in the UK by 1 April 1993. A reasonable level for humans perhaps, but is it safe for plants?

The problems don't stop there, though. Much of the SO_2 in the air will be oxidised, ultimately dissolving in rain to form dilute H_2SO_4. 'Normal' rain has a pH of about 5.6 (because of the dissolved carbon dioxide). Rainfall in industrial areas often has a pH of 4.5 or less, and the most acidic rain on record had a pH of 1.5. Acid rain attacks metal and stone, and lowers the pH of lakes and rivers. Studies of the lakes in

CONTINUED

the USA have shown that their pH often fell below 5. When this happened, metal ions – particularly Al^{3+}, Mn^{2+}, Zn^{2+} and Fe^{3+} – were released into the water from sediments. Aluminium is particularly toxic to fish. By the time the pH had reached 4.5, virtually all the fish were dead.

What can be done? The complete removal of the SO_2 from waste gases is just not possible, but substantial reductions can be achieved. One process involves passing the waste gases through a wet slurry of lime, or through a fluidised bed of calcium or sodium carbonate powder. Alternatively, it can be reduced to hydrogen sulphide (H_2S) with methane or coal, followed by catalytic conversion to sulphur. The world production of sulphuric acid in 1987 used 51 million tonnes of sulphur. Compare this with the 200 million tonnes blown up chimneys! By 1985 the USA had 87 flue gas desulphurisation plants in operation, Japan had 110 and West Germany, 13. Under pressure from EEC directives, Great Britain plans to have several in operation by the mid-1990s.

Sulphur has other polluting effects too. The foul smell of stagnant ponds or blocked drains arises from sulphur compounds. Bacterial decomposition uses up the dissolved oxygen in the water, after which only anaerobic decay can occur, producing hydrogen sulphide, dimethylsulphide ($(CH_3)_2S$), and other sulphur compounds. Using gas chromatography, sulphur compounds have also been identified in bad breath (dimethylsulphide and dimethyldisulphide, $(CH_3)_2S_2$) and in the odour of garlic (diallyl disulphide, $(CH_2=CH-CH_2)_2S_2$ and related compounds). Not all sulphur compounds smell nasty; one is responsible for the delicious aroma of fresh coffee.

Another source of bad smells is flatus. Although merely tricky in polite circles, this is a serious problem for NASA. Imagine being cooped up with this gas in a spacecraft for several days. In addition, flatus contains H_2S, a weak acid, which could corrode vital components. So the early American spacecraft carried catalytic converters to oxidise this sulphur to SO_2 which could be dissolved in alkali.

To maintain life-support systems in space we have to control our pollutants. How long will it be before the life-support systems of Spaceship Earth force us to do the same? In 1983 Norway proposed that within ten years the industrialised countries of the West should aim to cut SO_2 emissions to 30% of their level in 1980. Over 40 nations have so far signed and ratified this treaty.

Questions

Use the relative atomic masses from your data book or periodic table to answer these questions.

1 mole of any gas at *room* temperature and pressure occupies about $24\,dm^3$ or $24\,000\,cm^3$

$$1 \text{ tonne} = 10^3 \text{ kg} = 10^6 \text{ g}$$

$$1 \text{ m}^3 = 10^3 \text{ dm}^3 = 10^6 \text{ cm}^3$$

1. Use table 1 to answer this question.

   (a) Why do you suppose data was collected from site 4?

   (b) Give two possible reasons why the SO_2 concentrations were higher at site 1 than at site 4.

   (c) What does the data suggest about the effect of SO_2 on trees? Quote the data you refer to.

   (d) The data does not *prove* that SO_2 affects trees. Why not?

   (e) Suggest an experiment you could set up to *prove* whether or not SO_2 affects trees.

CONTINUED

Sand castles and mud huts © 1991 Jeffrey Hancock, published by Hodder and Stoughton Educational

2. In 1983, about 847 000 tonnes of sulphur fell on the United Kingdom.

(a) This sulphur originally entered the atmosphere as SO_2. What mass of SO_2 was involved?

(b) What was the volume of the SO_2 (measured at room temperature and pressure)?

(c) If we assume that all this SO_2 was oxidised and came down in rain as H_2SO_4, what mass of H_2SO_4 was deposited?

3. Sulphur dioxide concentrations are expressed either as parts per million by volume (ppm), or as micrograms per cubic metre ($\mu g\,m^{-3}$). If a sample of air contained one ppm of SO_2 (measured at room temperature and pressure):

(a) what volume of SO_2 would there be in one cubic metre?
(b) what mass of SO_2 would there be in micrograms per cubic metre?

4. Sulphur dioxide irritates human lungs because it is so soluble in water. When breathed in, the SO_2 dissolves in the mucous membranes, forming sulphurous acid, H_2SO_3. (Oxidation to H_2SO_4 may occur too.) This may cause scarring and eventually lead to chronic lung disease. The SO_2 level at Templemore Road, Belfast, in 1988 averaged 15.7 parts per million. If the volume of air inhaled each time an average person breathes quietly is about 500 cm^3 (more during exercise), use this information to estimate the following.

(a) The total volume of air one person inhaled during 1988. (Estimate how often you breathe.)

(b) The volume of SO_2 breathed in.

(c) The number of moles of sulphurous (or sulphuric) acid formed in her lungs.

5. Ferrybridge is a 2000 MW coal-fired power station near Leeds. PowerGen plc intend to install a flue gas desulphurisation (FGD) plant to remove 90% of the SO_2 from its waste gas. In this plant the SO_2 reacts with calcium carbonate ($CaCO_3$) powder, in an air blast to ensure oxidation, to produce the sulphate:

$$2CaCO_3 + 2SO_2 + O_2 \rightarrow 2CaSO_4 + 2CO_2$$

This plant can use up to 10 000 tonnes of limestone a week. Calculate:

(a) the mass of SO_2 that can be removed from the flue gas per week,

(b) the total mass of SO_2 produced by the power station each week,

(c) the total mass of sulphur burnt in the power station each week,

(d) the percentage of sulphur in the coal, if 1.2×10^5 tonnes of coal are burnt each week,

(e) the mass of gypsum, $CaSO_4 . 2H_2O$, that will be produced by this process, (PowerGen plc plan to dispose of this by selling it to make plasterboard and by adding it to the ash that they dump for landfill).

(f) the mass of CO_2 produced by burning coal, (assume that the coal is just carbon).

(g) the mass of CO_2 produced by the FGD process. CO_2 is a greenhouse gas: will the FGD process have a significant effect on the amount of CO_2 produced by the power station?

6. Raffinose ($C_{18}H_{32}O_{16}$) is a sugar that occurs in pulses, such as beans. It is poorly digested by our enzymes, so it arrives at the large intestine essentially intact. There it is attacked by bacteria. The products vary

CONTINUED

widely (and include sulphur compounds – see question 7), but one possibility might be:

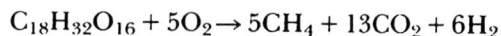

$$C_{18}H_{32}O_{16} + 5O_2 \rightarrow 5CH_4 + 13CO_2 + 6H_2$$

If raffinose constitutes 0.1% of a 500 g can of baked beans, estimate the total volume of the products.

7. A NASA publication has suggested that if all the sulphur compounds produced by an astronaut (principally H_2S from flatus) were oxidised and dissolved in water, about 2 g of H_2SO_3 would be produced per day. What volume of H_2S do they expect an astronaut to produce?

8. There are many methods of determining the SO_2 levels in air. One involves sucking a measured volume of air through hydrogen peroxide:

$$H_2O_2 + SO_2 \rightarrow H_2SO_4$$

Then the sulphuric acid is titrated with a standard solution of an alkali. On 7 December 1952, 875 people were killed by the smog. At an air sampling station in central London, 1.95 m^3 of air were drawn through hydrogen peroxide. The resulting H_2SO_4 required 11.70 cm^3 of Na_2CO_3 solution of concentration 5.00×10^{-3} mol dm^{-3} (5.00×10^{-3} M) for neutralisation:

$$Na_2CO_3 + H_2SO_4 \rightarrow Na_2SO_4 + H_2O + CO_2$$

Calculate:

(a) the number of moles of acid, and hence of SO_2 produced,

(b) the mass of SO_2 in 1.95 m^3 of air,

(c) the concentration of SO_2 in air in (i) micrograms per cubic metre, and in (ii) parts per million. (Use the conversion factor from question 3.)

Compare this figure with the EEC limit mentioned in the text. There has been progress!

9. Table 2 contains data recorded at sampling stations at various times. In each case, calculate the SO_2 concentration in micrograms per cubic metre.

TABLE 2

Site	Date	Volume of air/m^3	Volume of 5.00×10^{-3} M Na_2CO_3 solution/cm^3
Moor Lane, London	Dec 1970	5.40	16.10
Moor Lane, London	Nov 1988	9.25	2.90
Cox's Lane, Mansfield	Nov 1988	12.46	13.90

10. Sulphur dioxide can also be measured by dissolving it in water and titrating it with acidified potassium dichromate:

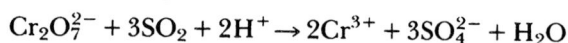

$$Cr_2O_7^{2-} + 3SO_2 + 2H^+ \rightarrow 2Cr^{3+} + 3SO_4^{2-} + H_2O$$

One day in Oxford in November 1988, 11.20 m^3 of air were drawn through water, and the resulting solution required 5.70 cm^3 of 2.00×10^{-3} mol dm^{-3} potassium dichromate for reaction. Calculate the concentration of SO_2 in micrograms per cubic metre.

Sand castles and mud huts © 1991 Jeffrey Hancock, published by Hodder and Stoughton Educational

10. Taking drugs

What happens when you swallow any medicine? First, the drug is absorbed through the wall of the stomach or the gut, and passes into the bloodstream. This carries the drug around the body, to wherever it is to act. As soon as it has been absorbed, however, a variety of chemical reactions start, catalysed by the body's enzymes, converting it into different substances. Thus the concentration of the drug in the bloodstream will fall, eventually to zero.

This is both an advantage, and a problem. It is useful in that the drug is needed only while you are ill. Prolonged exposure to the drug may be harmful, so it is convenient that the body acts to get rid of it. On the other hand, if the concentration of a drug is continually falling, its effectiveness will fall, too.

The data in table 1 refer to the antibiotic penicillin G. A man of mass 70 kg (about 154 lb or 11 stone) was given a 500 mg penicillin tablet and at the time intervals indicated, samples of his blood were withdrawn and analysed.

TABLE 1

Time/minutes	5	10	20	35	45	60	90	120
Penicillin G concentration/ mg dm^{-3} of blood	8.8	7.7	5.9	4.0	3.0	2.0	0.9	0.4

Questions

On rates of reactions

1. The rate of breakdown of the penicillin could be represented by the rate equation:

$$\text{Rate} = k[\text{penicillin}]^n$$

where [penicillin] = concentration of penicillin in mol dm^{-3}

k = reaction rate (velocity) constant

and n = a constant.

(a) What is meant by:

(i) the reaction rate (velocity) constant?

(ii) the order of a reaction?

(b) What would the rate equations be if the reaction was:

(i) first order in penicillin?

(ii) second order in penicillin?

2. (a) Plot the data with the penicillin concentration on the y-axis. Join the points with a smooth curve.

(b) What is the half-life of a reaction?

(c) Determine how long it takes for the penicillin concentration to fall from:

(i) 8.0 to 4.0 mg dm^{-3}

CONTINUED

     (ii) 4.0 to 2.0 mg dm^{-3}

     (iii) 2.0 to 1.0 mg dm^{-3}

     (iv) 1.0 to 0.5 mg dm^{-3}.

  (d) What is the half-life of penicillin G in the body?

  (e) What is the order of the decay reaction? Explain your answer.

  (f) Give the rate equation for the breakdown of penicillin G.

  (g) What are the units of the reaction rate constant?

  (h) Calculate a value for the rate constant.

3. (a) Extrapolate the graph from question 2(a) to find the concentration of penicillin G at zero time, assuming that it was all absorbed instantly. (It won't be, of course.)

  (b) Use this to find the total volume of liquid into which the penicillin was absorbed.

    These two figures, the half-life and the volume of the liquid into which the drug was absorbed, are important in drug therapy, and show a wide range of values. For example, half-lives may range from a quarter of an hour for aspirin to 8.9 ± 3.1 days for the anti-malaria drug chloroquine. If a drug has a short half-life, its concentration will rapidly fall to a level at which it is no longer effective, and the dose will have to be repeated.

    The volume of liquid into which the drug is absorbed (called the distribution volume) will also vary. This is because there is a wide range of fluid types in the body and different drugs will dissolve to varying extents in these fluids. Our 70 kg man will have about 5 dm^3 of blood, 20 dm^3 of liquid in the body's cells and 10 dm^3 of other fluids elsewhere, neither in the cells nor in the blood. Many drugs will have some fat solubility too, further increasing the range of distribution volumes.

    Virtually all drugs have some harmful side-effects – for example, aspirin can cause stomach bleeding – and the higher the concentration of the drug, the greater will be the side-effects. The challenge in drug treatment involves using a high enough concentration of the drug to achieve the desired result, but not such a high concentration that the side-effects become serious.

Questions

4. (a) The drug theophylline is a bronchodilator. It opens the air passages, and is thus useful against asthma. It has a half-life in the body of eight hours. A man took a 1020 mg tablet of theophylline. It dissolved in 35 dm^3 of body fluid. Calculate the concentration of theophylline (in mg dm^{-3}):

    (i) immediately after the tablet was taken, (assume that it was instantly absorbed),

    (ii) eight hours after it was taken, (remember that the half-life of theophylline in the body is eight hours),

    (iii) sixteen hours after it was taken,

    (iv) twenty-four hours after it was taken.

  (b) A *second* 1020 mg tablet was swallowed, 24 hours after the original one.

    (i) Calculate the *new* theophylline concentration. (Remember to take into account what was left of the first tablet.)

CONTINUED

(ii) Carry on calculating the concentration of theophylline at eight hour intervals for another 24 hours, that is, 48 hours after the first tablet.

(c) *Sketch* the graph of theophylline concentration against time. Plot the concentration of theophylline (in mg dm^{-3}) on the y-axis, and time on the x-axis.

5. Instead of taking one 1020 mg tablet every 24 hours, the man took one 340 mg tablet every eight hours. Use the method of question 4 to get a sketch graph of theophylline concentration against time for a total of 48 hours. (Find the concentration every 8 hours, as you did in question 4.)

6. The major problem with theophylline is the narrow margin between the therapeutic dose and the toxic dose. The therapeutic dose (the dose required for the drug to ease the asthma attack) is between 10 and 20 mg dm^{-3}. Above 20 mg dm^{-3}, toxic side-effects, such as nausea, become serious. Inspect your sketch graphs for questions 4 and 5 and suggest which is better, one 1020 mg tablet every 24 hours or one 340 mg tablet every eight hours. Explain your answer.

7. Antibiotics – like the penicillins – typically are taken every few hours, while chloroquine is taken once a week. Explain why.

Alcohol (ethanol) behaves in a similar way to other drugs. It is absorbed through the stomach and gut walls, and 95% of it is broken down in the liver. (The other 5% of it is lost in exhaled air and urine.) Unusually, however, once the alcohol has reached the liver, it is removed *at a constant rate*. This rate varies from person to person. Values as low as 4.1 g per hour and as high as 11.1 g per hour have been found. The *average* rate of breakdown of the ethanol in the liver is about 7.3 g per hour for a man, and 5.3 g per hour for a woman.

Despite much work, no way has been found to speed up the breakdown of alcohol in the body. Drugs, caffeine, exercise, fresh air – all have no effect. Getting used to alcohol doesn't help either; regular drinkers don't sober up more quickly.

Questions

8. A pint of beer (4% ethanol by volume) contains about 16 g ethanol.

(a) The distribution volume for ethanol for a 70 kg (11 stone) man is 42 dm^3. What will be the concentration of ethanol in his body fluids after he has drunk one pint?

(b) How many pints can he drink before he reaches the legal limit for driving (80 mg per 100 cm^3 blood)?

(c) The distribution volume for his 51 kg (8 stone) wife is about 22 dm^3. How many pints can she drink and still drive legally?

9. (a) If a man drinks five pints of beer or five double whiskies, he consumes about 80 g ethanol. *Sketch* a graph showing the amount of alcohol in his body against time, assuming that his liver removes it at a constant rate of 7.3 g per hour. How long will it take for all the ethanol to be removed? If he drank all this in the hour before closing time at 10.30 p.m., at what time will his body be ethanol free?

(b) What is the order of a process of this type? Write a rate equation for the process.

(c) Calculate or use your sketch from (a) to find how long it would take for the amount of ethanol to fall:

(i) from 80 g to 40 g,

CONTINUED

(ii) from 40 g to 20 g,

(iii) from 20 g to 10 g,

(iv) from 10 g to 5 g.

(d) What can you say about the half-life of this process?

10. In a court case in 1988, a driver involved in a fatal road accident gave a blood sample to the police four hours after the accident. The sample contained 41 mg of ethanol per 100 cm³ of blood.

(a) What was the total amount of ethanol in the driver's body at the time of his blood test? (Assume a distribution volume of 42 dm³.)

(b) Assuming that his liver removed the ethanol at the *average* rate for a man, what was the total amount of ethanol in his body at the time of the accident?

(c) What was his blood alcohol content at the time of the accident (in mg per 100 cm³)?

(d) The driver was convicted of driving with excess alcohol in his blood. But the British Medical Association has strongly advised against this back calculation to determine an alcohol concentration some hours before. In Sweden, the courts will not accept that this evidence is valid. Why not?

11. To provide a simple method of calculating alcohol intake, it is assumed that a 'unit of alcohol' is equivalent to 10 cm³ of ethanol, and that the liver can break down one unit of ethanol an hour. Is this a good estimate? (The density of ethanol is 0.79 g cm⁻³.)

12. Why should the kinetics of removal of penicillin G and of alcohol be so different? (*Hint*: Calculate the concentration of each reagent in the blood, in mol dm⁻³. Penicillin G has the formula $C_{16}H_{18}N_2O_4S$. What do you suppose will be the concentration of alcohol dehydrogenase in the liver?)

Sand castles and mud huts © 1991 Jeffrey Hancock, published by Hodder and Stoughton Educational

11. In the balance

Questions

On equilibria

1. (a) Certain centipedes produce a dilute aqueous solution of hydrocyanic acid, which dissociates in aqueous solution:

$$HCN \rightleftharpoons H^+ + CN^-$$

If the concentration of HCN is 0.1 mol dm^{-3}, what is the concentration of cyanide ions? [$K_c = 4.9 \times 10^{-10}$ mol dm^{-3}.]

(b) Cyanide ions are poisonous because they complex with Fe^{2+} and Fe^{3+} ions in haemoglobin and in enzymes. How strongly do they complex? The equilibrium is:

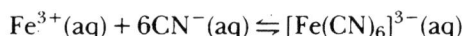

$$Fe^{3+}(aq) + 6CN^-(aq) \rightleftharpoons [Fe(CN)_6]^{3-}(aq)$$

Write an expression for the equilibrium constant (K_c). Calculate a value for K_c if the concentrations of the ions at equilibrium in a blood sample were found to be:

$$Fe^{3+} = 1 \times 10^{-6}, CN^- = 3 \times 10^{-6}$$

and $$[Fe(CN)_6]^{3-} = 1 \times 10^{-8} \text{ mol dm}^{-3}$$

2. Biological membranes consist largely of lipids (non-polar covalent compounds). If a drug is to be absorbed through a membrane, it must be able to dissolve in a non-polar medium, so it should be *uncharged*. Morphine is a strong painkiller and has a tertiary amine group in the molecule (marked in **bold** in fig 1).
Like all tertiary amines it is a base:

$$R_3N + H_2O \rightleftharpoons R_3NH^+ + OH^-$$

The equilibrium constant for this equilibrium (K_b) is given by the expression:

$$K_b = \frac{[R_3NH^+][OH^-]}{[R_3N]}$$

$K_b = 2.5 \times 10^{-6}$ mol dm^{-3}. Calculate values for the ratio:

$$\frac{\text{concentration of morphine (R}_3\text{N)}}{\text{concentration of protonated morphine (R}_3\text{NH}^+)}$$

(a) in the stomach, where the concentration of OH$^-$ ions is 3.2×10^{-13} mol dm^{-3},

(b) in the blood, where the concentration of OH$^-$ ions is 4.0×10^{-7} mol dm^{-3}.

(c) Is morphine best administered by mouth or by injection?

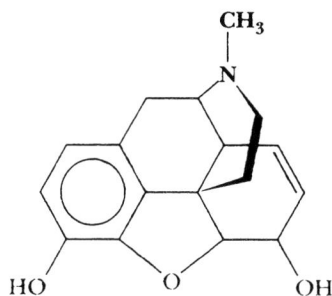

FIG 1

CONTINUED

3. Copper compounds are used to kill fungi and algae but are toxic to humans. After an industrial accident, a worker was admitted to hospital with copper poisoning. One possible treatment was to inject her with a solution of EDTA, which forms a complex with the copper ions:

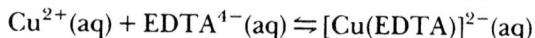

$$Cu^{2+}(aq) + EDTA^{4-}(aq) \rightleftharpoons [Cu(EDTA)]^{2-}(aq)$$

The copper complex would be excreted in the urine.

(a) Write down the expression for the equilibrium constant.

(b) If the concentration of free $EDTA^{4-}$ ions in the blood can be raised to 1×10^{-8} mol dm^{-3}, and the worker excretes the copper complex at a concentration of 1.5×10^{-8} mol dm^{-3}, estimate a value for the concentration of free copper ions in her body.
[$K_c = 1.6 \times 10^{18}$ mol^{-1} dm^3.]

(c) A problem with this treatment is that it also removes essential ions, such as Ca^{2+}. There is an alternative treatment, with a chemical called BAL (British anti-lewisite, $CH_2SH-CHSH-CH_2OH$, devised as an antidote to the chemical weapon lewisite). Table 1 gives approximate values for the equilibrium constants for complex formation.

Which treatment would you advise? Explain.

TABLE 1

	Equilibrium constant for complex formation with	
	BAL	EDTA
Ca^{2+}	6×10^1	1.0×10^{11}
Cu^{2+}	6×10^{22}	1.6×10^{18}

4. The recycling of old car batteries inevitably produces a lot of waste water containing toxic Pb^{2+} ions. A company wants to discharge this water into a river. One possible way to remove the Pb^{2+} ions would be to add aqueous NaCl, which sets up the equilibrium:

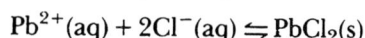

$$Pb^{2+}(aq) + 2Cl^-(aq) \rightleftharpoons PbCl_2(s)$$

for which the equilibrium constant, $K = [Pb^{2+}][Cl^-]^2$ and has a value of 2.0×10^{-5} mol^3 dm^{-9}. (For an explanation of this, see the section on solubility products in any Physical Chemistry textbook.)

(a) What concentration of Cl^- ions would the manufacturer have to use to lower the Pb^{2+} ion concentration to:

(i) 1×10^{-2} mol dm^{-3}?

(ii) 1×10^{-4} mol dm^{-3}?

(iii) 1×10^{-6} mol dm^{-3}?

(iv) zero?

(b) Is it ever feasible to remove all of a pollutant? Explain your answer.

5. The acidity of blood is maintained within a very narrow range by a mixture of H_2CO_3 and HCO_3^- ions, which exist in the equilibrium:

$$H_2CO_3(aq) \rightleftharpoons HCO_3^-(aq) + H^+(aq)$$

(a) What would be the effect on this equilibrium of addition of acid? Explain.

(b) Why would this stop the blood becoming too acidic?

(c) And what about the addition of alkali; what would the effect of that be?

(d) The H_2CO_3 is also in equilibrium with CO_2:

$$H_2CO_3(aq) \rightleftharpoons H_2O(l) + CO_2(g)$$

Anything which causes you to breathe more rapidly – panic, for example – makes you exhale unusually large amounts of CO_2. Explain why panic causes the pH of your blood to rise.

CONTINUED

Sand castles and mud huts © *1991 Jeffrey Hancock, published by Hodder and Stoughton Educational*

6. The reaction of nitrogen and oxygen is an equilibrium:

$$N_2 + O_2 \rightleftharpoons 2NO \qquad \Delta H = +181 \text{ kJ mol}^{-1}$$

Nitrogen monoxide (NO) is formed during the combustion of fuels in car and aircraft engines. Diesel engines operate at a higher temperature than petrol engines. What effect will this have on the amount of NO in the exhaust gases?

7. Nitrogen monoxide (NO) and carbon monoxide (CO) react according to the equilibrium:

$$2CO + 2NO \rightleftharpoons 2CO_2 + N_2$$

(a) Write an expression for the equilibrium constant for this equilibrium, K_p.

(b) The equilibrium constant is over $10^{100} \text{ kPa}^{-1}$ at the temperature of a car exhaust. What does this tell you about the amounts of CO, NO, CO_2 and N_2 you would expect to find in the car exhaust?

(c) Car exhaust gases contain significant quantities of both CO and NO. Explain why.

(d) If the exhaust gases are passed through a fine platinum gauze, the amounts of CO and NO become negligible. Why is this?

12. Tooth decay

As we get older, we decline: our skin wrinkles, hair thins, breasts sag – and our teeth decay. It wasn't always like that. Bronze Age men may have been bald and lined, but studies of their skulls suggest that only about 2% of their teeth were decayed. An examination of Eskimos' teeth confirms this; only when they start to eat a modern diet do their teeth rot. And *our* teeth rot in large numbers. A survey of British school children some years ago showed that on average, each child had 5.5 decayed, filled or missing teeth.

What is it that rots our teeth so effectively? Sugar!

For example, look at the tooth decay of nearly 8000 Japanese schoolchildren as their sugar consumption varied during and after the Second World War (fig 1). Rats fed on a high sugar diet suffer from tooth decay too, but not if the sugar is passed directly to their stomach using a tube.

How can sugar cause this decay? Teeth collect a substance called plaque on their surface. Normally the pH of this plaque is close to 7, but if food is eaten, it falls rapidly below 4. (Curve (a) in fig 2.)

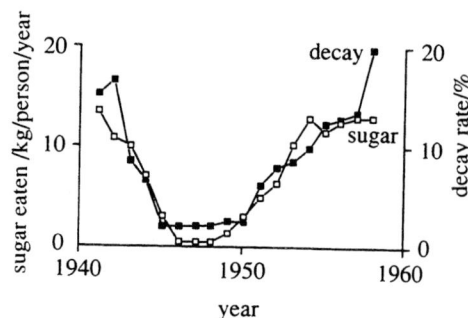

FIG 1

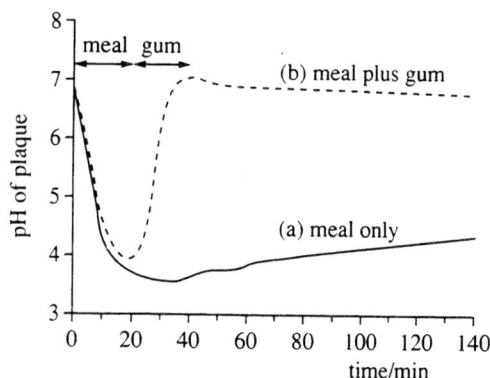

FIG 2

The sugar is not responsible for this fall in pH because sugars are not acidic. It has been suggested that bacteria in plaque break down the sugars in food to produce a mixture of acids, including 2-hydroxypropanoic acid (lactic acid). In support of this, it is known that if rats are bred in a sterile environment so that their mouths are free of bacteria, they have little tooth decay, however much sugar they eat.

What a chef would call sugar, we should more properly call sucrose, and although we recognise this as a damaging compound, it seems that any natural sugar can be converted by bacteria into dangerous acids.

Questions

On acids and acidity theory

1. (a) Define pH.

(b) Suggest *two* ways by which the plaque's pH might have been measured to get the data plotted in fig 2.

CONTINUED

Sand castles and mud huts © 1991 Jeffrey Hancock, published by Hodder and Stoughton Educational

2. (a) Does fig 1 suggest that sugar consumption and tooth decay are linked? Explain your answer.

(b) Explain why fig 1 does not *prove* that sugar causes tooth decay.

(c) A sugar manufacturer argues that tooth decay and sugar consumption are unrelated. Suggest an argument he could use to explain fig 1.

3. (a) 2-hydroxypropanoic acid (lactic acid) ionises as follows:

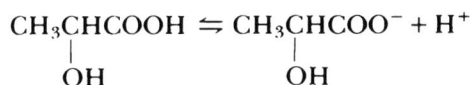

$$CH_3CHCOOH \rightleftharpoons CH_3CHCOO^- + H^+$$
$$\quad\;\, | \qquad\qquad\qquad | $$
$$\quad\; OH \qquad\qquad\quad\; OH$$

Write an expression for the dissociation constant, K_a.

(b) After a meal, the plaque's pH falls to 3.7. Calculate the hydrogen ion concentration under these conditions.

(c) If this acidity is assumed to arise solely from the presence of 2-hydroxypropanoic acid, what is the concentration of the *undissociated* acid? ($K_a = 1.3 \times 10^{-4}$.)

(d) Sodium lactate (sodium 2-hydroxypropanoate, E325) is added to confectionery and cheese. As a result, eating these products increases the concentration of lactate ions in the mouth. What effect will this have on the plaque's pH? Explain. (No calculations are required.)

4. It has been suggested that plaque's acidity is not due to a simple carboxylic acid but to an amino acid. Do you think this is possible? What concentration of amino acid would be needed to reach a pH of 3.7? (The pK_a for glycine, the simplest amino acid, is 9.87.)

5. A health-conscious man breakfasted on grapefruit, which contains citric acid, at a concentration of $0.100 \text{ mol dm}^{-3}$. Citric acid is a tribasic acid ($C_3H_5O(COOH)_3$). For simplicity, let us call it $CitH_3$.

(a) If K_1 is the dissociation constant for the ionisation of the first proton, that is, for the equilibrium:

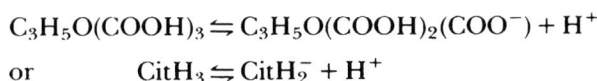

$$C_3H_5O(COOH)_3 \rightleftharpoons C_3H_5O(COOH)_2(COO^-) + H^+$$

or $\qquad CitH_3 \rightleftharpoons CitH_2^- + H^+$

K_2 is the constant for dissociation of the second proton:

$$CitH_2^- \rightleftharpoons CitH^{2-} + H^+$$

and so on, write down expressions for K_1, K_2 and K_3.

(b) K is the overall dissociation constant, that is, the constant for the dissociation of *all* the protons:

$$CitH_3 \rightleftharpoons Cit^{3-} + 3H^+$$

Write down an expression for K.

(c) Show that $K = K_1 \times K_2 \times K_3$, and hence that $pK = pK_1 + pK_2 + pK_3$.

(d) If $pK_1 = 3.14$, $pK_2 = 4.77$ and $pK_3 = 6.39$, use the expressions in (c) to calculate the overall pK and hence a value for K for citric acid.

(e) Calculate the pH of the man's mouth. [*Hint*: Assume that x moles dm^{-3} of the acid had dissociated, and find the concentrations of H^+ and Cit^{3-} ions produced.]

Tooth enamel is hydroxyapatite ($Ca_5(PO_4)_3OH$). When it is exposed to acid, two things happen. In a slightly acid solution, some Ca^{2+} ions dissolve:

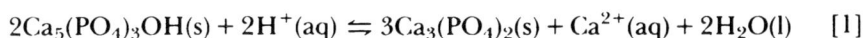

$$2Ca_5(PO_4)_3OH(s) + 2H^+(aq) \rightleftharpoons 3Ca_3(PO_4)_2(s) + Ca^{2+}(aq) + 2H_2O(l) \qquad [1]$$

CONTINUED

If the pH falls further, all the Ca^{2+} ions dissolve:

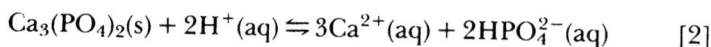

$$Ca_3(PO_4)_2(s) + 2H^+(aq) \rightleftharpoons 3Ca^{2+}(aq) + 2HPO_4^{2-}(aq) \qquad [2]$$

Fortunately plaque's pH gradually returns towards 7 under the influence of saliva. This is a complex liquid, with more functions than just keeping your mouth wet. Firstly, it acts as a buffer, holding the pH of your mouth close to 7. A variety of systems help to do this, but the main buffering action is done by the hydrogen carbonate/carbonic acid system.

Saliva's second major function is to allow the remineralisation of teeth. When the pH rises, equations [1] and [2] are reversed. Calcium, phosphate and hydroxide ions from the saliva are redeposited, reforming the enamel of the tooth.

Knowing all this, what are we to do to prevent decay?

We can avoid the food that causes it, especially sucrose. A famous investigation in Sweden between 1945 and 1953 showed that you do not have to avoid sucrose entirely, but that the important thing is to eat only at meal times. Sugary snacks between meals have the worst effect on teeth.

The second thing to do is to remove plaque and tooth debris from the teeth, most obviously by brushing. Curve (b) in fig 2 illustrates another method. It plots the plaque's pH when a meal was followed by chewing sugar-free gum for 20 minutes. The gum also stimulates the flow of saliva.

Finally one can attempt to alter the material of the tooth itself. The most spectacular success has been achieved by the use of fluoride, either in toothpastes or added to the water supply. The fluoride ions react with the hydroxyapatite, converting it to the less soluble fluorapatite, $Ca_5(PO_4)_3F$. Addition of one part per million of fluoride ions (F^-) to the water reduces the rate of decay by half. No ill-effects can be detected even at levels of eight parts per million.

Questions

6. Saliva is a complex mixture, with a pH of about 7.0.

(a) One of its components is ammonia, formed by the breakdown of proteins in food. Ammonia is useful in that it reacts with mouth acids. Write an equation to illustrate this.

(b) Saliva also contains various buffer systems. What is a buffer?

(c) In stimulated saliva (for example, when eating), this buffering is due largely to the equilibrium:

$$H_2CO_3 \rightleftharpoons HCO_3^- + H^+$$

Explain how this buffer system works.

(d) The concentration of HCO_3^- ions is 0.01 mol dm^{-3}. What is the concentration of H_2CO_3? [pK_1 for carbonic acid is 6.4.]

(e) There are other buffer systems present in saliva. One of them is similar to the system involving carbonate ions above, but involves phosphate ions (PO_4^{3-}). Write down the equilibria for this system.

7. Suggest two reasons why chewing sugar-free gum can help prevent tooth decay.

8. Toothpastes are complex mixtures of compounds.

(a) They must not contain compounds such as glucose, lactose or sucrose. Why not?

(b) There is always an abrasive present, to scour the plaque off the teeth. Early abrasives were compounds of Ca^{2+} and PO_4^{3-} ions,

CONTINUED

Sand castles and mud huts © *1991 Jeffrey Hancock, published by Hodder and Stoughton Educational*

such as $CaHPO_4 . 2H_2O$. Use equations [1] and [2] to suggest another advantage of these compounds in addition to their abrasive qualities.

(c)   Toothpastes often contain an ionic fluoride, because even with this short contact time, some fluorapatite can be formed. Unfortunately, F^- ions react with H^+ ions:

$$F^- + H^+ \rightleftharpoons HF$$

Of two fluoride toothpastes tested, one had had its pH adjusted to 6.0, the other to pH 8.0. Which contained the higher concentration of F^- ions? Explain your answer.

9.   (a)   It is known that about 5 g of sodium fluoride are enough to kill an average person. If a drinking water supply contains one part per million of fluoride ions, what volume of water would an individual have to drink to consume a lethal dose of fluoride?

(b)   One part per million of fluoride ions in the water supply about halves the rate of tooth decay. Why is it not universally adopted? (In the UK only about 10% of the population has a fluoridated water supply.) Suggest some arguments that an opponent of fluoridation might use.

10.   Because saliva is buffered, it can be titrated with acid. Table 1 gives the volume of 0.02 M HCl that had to be added to 100 cm³ of saliva to reduce its pH to 5.5. It is divided into two groups, one using saliva from eleven children who had no tooth decay, and the other using saliva from another eleven children with a lot of decay.

TABLE 1

Children with:	Volume of 0.02 M HCl for 100 cm³ of saliva										
no decay	134	123	94	95	101	108	110	111	115	116	118
much decay	141	108	94	106	96	107	93	91	90	85	75

What conclusions can you draw? (The size of this sample is not large enough for firm conclusions, but I have selected this data from a much larger sample, which confirmed the general picture.)

13. Blood, sweat and seas

Sea water is salty. This is because rain water dissolves small amounts of rocks as it trickles through them, and rivers carry these dissolved solids to the sea. Evaporation then removes pure water, and so the process goes on, increasing the salt concentration of the sea water. Or so it was assumed. The trouble is that sea water contains about 1.1% of sodium ions, for example, and it would have taken the rivers only about 75 million years to produce this. But we know that the oceans are at least 40 times older than that, so the salt concentration should be higher.

Table 1 gives the amounts of the elements calcium, sodium and potassium in the Earth's crust and in sea and river water.

There are several puzzles. For example, why is potassium so much less abundant than sodium in river and sea water, even though it is almost as widespread in the Earth's crust? Perhaps this is because naturally occurring potassium compounds are less soluble in water than naturally occurring sodium compounds.

That cannot be true for calcium compounds, though. There is much more calcium than sodium in the Earth's crust and in river water. Why is there so little of it in the sea? We must assume that there are other processes which remove salts from sea water. Let us look at a few.

TABLE 1			
	Ca	Na	K
	/mg of the element per kg		
Earth's crust	46 600	22 700	18 400
Sea water	412	10 760	399
River water	13.4	5.2	1.3

Carbonate formation

Marine animals remove calcium ions from sea water to make their shells and skeletons from calcium carbonate. This alone explains the low calcium concentration in the sea. The animals remove more Ca^{2+} ions each year than all the rivers put in.

Evaporite formation

If a large amount of water evaporates from a body of sea water, the remaining salts become so concentrated that they crystallise out. Of course, this won't happen all the time – you need an enclosed area of sea, for example – but it happens periodically and is significant over geological time. (For example, the Mediterranean formed large quantities of these evaporites about six million years ago, when the link with the Atlantic through the Straits of Gibraltar was restricted.)

Ion exchange

Rivers also carry solid particles to the sea. The most important of these from our point of view are suspended particles of clay. These are complex aluminosilicates, with the gaps in their lattice mainly occupied by calcium ions. When these clays reach the sea, sodium and potassium ions can replace the calcium ions.

Questions

1. It was suggested in the passage that one reason why there is so little potassium in the sea is that potassium compounds are less soluble than sodium compounds. If this is true, when water flows through halite (NaCl) or carnallite ($KCl.MgCl_2.6H_2O$), the Na^+ ions will dissolve in greater quantities than the K^+ ions.

CONTINUED

Sand castles and mud huts © 1991 Jeffrey Hancock, published by Hodder and Stoughton Educational

(a) What sort of lattice do solid sodium and potassium chlorides have?

(b) Draw the lattice.

(c) When sodium chloride is dissolved in water, the Na^+ and Cl^- ions become hydrated. Explain what this means.

(d) Draw a diagram of these hydrated ions.

2. When sea water evaporates, it is the water that turns to vapour, and not the dissolved salts. Explain why.

3. Lattice enthalpies/$kJ\,mol^{-1}$ NaCl: 780 KCl: 711

Hydration enthalpies/$kJ\,mol^{-1}$ Na^+: -406 K^+: -322 Cl^-: -364

(a) Define lattice energy, enthalpy of hydration and enthalpy of solution.

(b) Why is the hydration enthalpy of the sodium ion greater than that of the potassium ion?

(c) Use the data above to calculate the enthalpies of solution of NaCl and KCl.

(d) KCl actually is less soluble in water than NaCl. Suggest why.

(e) Your data might lead you to expect that NaCl and KCl do not dissolve at all. Explain why.

(f) Why *do* KCl and NaCl dissolve in water?

4. When a calcium clay suspended in river water reaches the sea, it undergoes ion exchange. The Na^+ or K^+ ions in the sea water compete with the Ca^{2+} ions for the negatively charged sites on the clay:

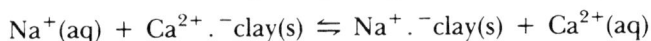

$$Na^+(aq) + Ca^{2+}.^-clay(s) \rightleftharpoons Na^+.^-clay(s) + Ca^{2+}(aq)$$

(a) We normally should expect the calcium ions to be attracted to the clay more strongly than sodium or potassium ions. Why?

(b) Why, then, do Na^+ ions replace the Ca^{2+} ions when the clay particles reach the sea?

(c) The Na^+ ion is smaller than the K^+ ion. Say why, and explain why this would lead us to expect that the Na^+ ions would be attracted more strongly to the clays than the K^+ ions.

TABLE 2

Ionic radii/nm	Li^+	Na^+	K^+
Bare, unhydrated ion	0.074	0.102	0.138
Hydrated ion*	0.34	0.28	0.23

(*This is the approximate radius of the ion *and the water*.)

(d) Why do the radii of the hydrated ions decrease down the group?

(e) Perhaps the ion remains *hydrated* when it interacts with the clay. Use this idea to explain why K^+ ions are attracted to clay particles more strongly than Na^+ ions.

TABLE 3

Concentrations	Na^+	K^+
	/mmol dm^{-3}	
Sea water	470	10
Blood plasma	145	5
Blood cells	10	150

Blood is salty, too. Mammals evolved in the sea, so their bodies' fluids might be expected to be fairly similar to it. The fluids *inside* the body cells have a very different composition, however, much higher in potassium ions and lower in sodium ions (table 3). This situation is

CONTINUED

maintained by an enzyme (called an Na^+K^+-ATPase), which pumps ions across cell membranes, moving K^+ ions in and Na^+ out.

The imbalance in Na^+ and K^+ concentrations between the inside and outside of the cells is crucial for the functioning of the body. For example, a nerve impulse starts when the wall of the nerve cell becomes more permeable to Na^+ ions, allowing them to flow down the concentration gradient into the cell. This is followed, about 5×10^{-1} seconds (0.5 ms) later, by K^+ ions moving out of the cell. These ion movements cause an electrical potential to be set up. The propagation of this potential along the nerve causes the impulse, and carries the message to the brain, to the muscles, and so on. About 5–10 ms after the impulse, the Na^+K^+-ATPase has restored the difference in Na^+ and K^+ concentrations, and the nerve is ready for action again.

Sweat is also salty. The loss of Na^+ ions in this and other ways means that we must replace them through our diet. But we must all beware because we *like* salt, and most people eat too much of it. Recent work indicates that this causes high blood pressure. In the UK we eat 8–10 g of NaCl a day, while the Indians of the Amazonian basin eat only 0.1 g. Our blood pressure rises as we get older (with an increased risk of strokes and heart failure), but that of the Amazonian Indians does not.

There is another interesting example of the effects of the Na^+/K^+ balance. People with manic-depressive illness suffer from periods of extreme elation alternating with deep, even suicidal, depression. No one knows what causes it, but it seems to arise from some chemical imbalance in the brain. Regular small doses of lithium salts have had great success in controlling a patient's mood swings, so maybe the problem lies in the Na^+/K^+ balance of the brain.

Questions

5. Biological cell walls can obviously distinguish Na^+ and K^+ ions very effectively. Our understanding of how this might happen was developed by the discovery of some organic compounds, called crown ethers.

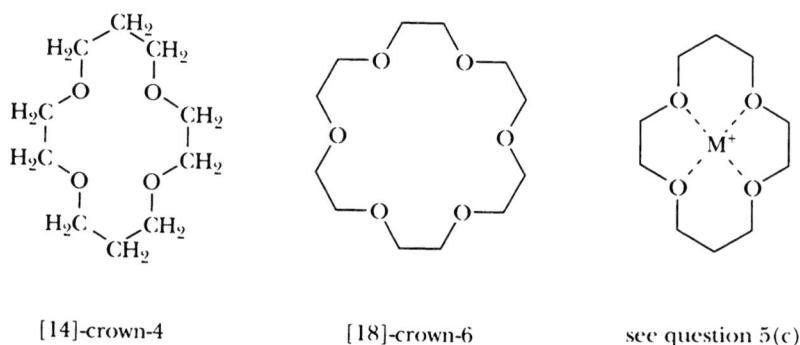

[14]-crown-4   [18]-crown-6   see question 5(c)

FIG 1

(a) Look at fig 1 and work out the logic behind the names. Draw the structure of [15]-crown-5.

(b) Salts of the Group I elements are not soluble in organic solvents like dichloromethane, CH_2Cl_2. Why not? (Refer back to your answer to question 3.)

(c) In the presence of crown ethers, however, they will dissolve in CH_2Cl_2. It is thought that the metal ion fits into the ring of the crown ether, interacting with the oxygen atoms (see fig 1). Explain how this helps the salt to dissolve in CH_2Cl_2.

CONTINUED

Sand castles and mud huts © 1991 Jeffrey Hancock, published by Hodder and Stoughton Educational

(d) Table 4 lists some data for the solubility of a series of Group I chlorides in CH_2Cl_2 in the presence of various crown ethers.

TABLE 4

| Crown ether added | Solubility in CH_2Cl_2 of chlorides of | | | | |
| | Li^+ | Na^+ | K^+ | Rb^+ | Cs^+ |
			$/10^{-5}$ mol dm^{-3}		
[14]-crown-4	2.1	1.6	0.8	0.6	0.5
[15]-crown-5	0.3	4.2	1.8	1.4	0.9
[18]-crown-6	0.2	1.8	5.4	4.1	3.1

(i) Which is the most soluble chloride in each case?

(ii) Can you explain why? (*Hint*: Refer back to table 2.)

6. (a) When Na^+ ions are pumped out of cells and K^+ ions are pumped in, they presumably are bonded to large molecules. Suggest how the molecules can distinguish between Na^+ and K^+ ions. [Use your answer to question 5(d).]

(b) Na^+K^+-ATPase also pumps Li^+ ions across cell walls, but not as efficiently as it pumps either Na^+ or K^+. Why doesn't it select Li^+ ions as well as Na^+ or K^+ ions?

14. Five per cent of us

Problems with aluminium

Aluminium makes up 8.3% of the Earth's crust, but is present in the human body in minute amounts. For example, the average person contains a *total* of only 35 mg of it. Evidence is beginning to accumulate that excess aluminium in the body has serious consequences.

In the early 1970s, patients on dialysis treatment for kidney failure began to show serious side-effects. The people so afflicted invariably died, often very rapidly. The problem was soon linked with the uptake of aluminium from the dialysis fluids. (It has now been eliminated.)

If aluminium causes problems in kidney patients, could it cause similar problems in the rest of the population? Alzheimer's disease has symptoms similar to those of the affected kidney patients. It may start quite early, and will affect 1 in 20 of everybody over the age of 70. When a person suffers from this, he or she becomes extremely confused and eventually incapable of sensible conversation or indeed living a normal life.

There are two strands of evidence that these symptoms are caused by aluminium. First there is the post-mortem evidence. The brains of sufferers from Alzheimer's disease have very characteristic tangled neurons. These tangles contain high aluminium concentrations. Secondly there is the statistical evidence: people who live in areas with a high concentration of aluminium in the water supply are more likely to suffer from the disease than people from areas where the aluminium concentration is low.

How much aluminium does an adult absorb? There has been much research done on this, and it seems that several factors may affect it.

The amount of aluminium in the water supply

Aluminium ions are very effective at coagulating suspensions, so aluminium sulphate is often added to water during the purification process, to precipitate out peat or other impurities. The pH of the water is then adjusted to reduce the solubility of the aluminium to a minimum. Fig 1 shows the variation in the solubility of aluminium with pH. (Note that the scale on the y-axis is not linear.)

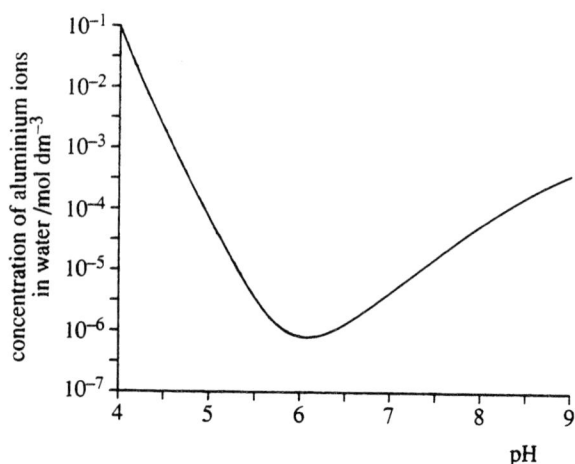

FIG 1

CONTINUED

Sand castles and mud huts © 1991 Jeffrey Hancock, published by Hodder and Stoughton Educational

The amount of aluminium in the food

Table 1 gives some data for the amounts of aluminium in various foods; (the dates refer to the date of publication).

TABLE 1

Food	Aluminium concentration/mg kg^{-1}					
	1980	1981	1985	1985	1987	1988
Carrots	3	5	26	–	–	0.05
Potatoes	2	1	8	2	0.2	<0.2
Apple	1	0.2	–	0.4	–	–
Tea	3	–	4	5	–	1–2
Cow's milk	0.7	0.9	–	0.7	–	0.02
Human's milk	0.5	–	–	0.3	–	–

There are problems associated with determinations of aluminium concentrations like these. For example, the values are rather small and so are difficult to determine. As techniques develop the values get more accurate, so the later figures are probably the most reliable.

Soil contamination

The concentration of aluminium in ordinary soil is about 50 000 mg per kg, so any contamination of food with soil will raise the food's aluminium concentration. On the other hand, much of the aluminium in soil is in the form of aluminosilicates, in which the aluminium is extremely strongly bonded, with no tendency to pass into solution. So, for example, although cows eat lots of earth (and aluminosilicates) with their grass, the aluminium levels in cows' milk are comparable to those in human milk. Apparently, the cows don't absorb much of the aluminium they eat.

Questions

1. Several water boards use $Al_2(SO_4)_3$ in the purification of the water supply.

(a) What should the pH of the water be adjusted to in order to precipitate out as much of the aluminium as possible?

(b) All of the aluminium cannot be removed. Why not?

(c) Use the data in the passage to calculate the concentration of the remaining aluminium in mg dm^{-3}.

2. The aluminium content of potatoes has been measured several times with widely differing results. Use the information in the table to suggest which is likely to be the best value. Explain your answer.

3. (a) Like all salts, those of aluminium undergo hydration when dissolved in water. Draw the structure of the resulting aluminium-containing species.

(b) Between pH 0 and 7 several hydrolysis equilibria are set up. What are these equilibria?

(c) Solutions of aluminium salts are acidic. Why?

(d) How will pH affect the hydrolysis equilibria?

CONTINUED

(e) A mixture of sodium aluminium phosphate and sodium hydrogen carbonate is used as a raising agent. When added to the moist cake mix, it produces CO_2 which causes the cake to rise. How does it work? Write equations for the reactions taking place.

4. When the pH of solutions of aluminium salts in water is raised above 7, other equilibria become significant.

(a) What are these?

(b) How will they be affected by pH?

5. (a) Explain the variation in solubility of aluminium ions in water at different pH values (fig 1) in the light of the equilibria in your answers to questions 3 and 4.

(b) As we shall see later, absorption of aluminium across the gut wall isn't a simple affair. We could represent it like this:

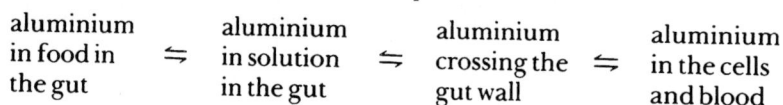

aluminium in food in the gut $\leftrightharpoons$ aluminium in solution in the gut $\leftrightharpoons$ aluminium crossing the gut wall $\leftrightharpoons$ aluminium in the cells and blood

If the aluminium isn't in *solution* in the gut it won't be absorbed. The pH of the stomach is about 1.5; that of the duodenum about 8. Will aluminium be absorbed from the stomach or the duodenum?

(c) The mineral diaspore $(AlO(OH))$ occurs in soils. Would you expect it to dissolve in the stomach (pH 1.5) or the duodenum (pH 8)? Explain your answer and write suitable equations.

Cooking or storage in aluminium

If food is cooked or stored in metal containers, some of the metal will dissolve in the food. The pH of the food presumably will affect how much metal is absorbed. Fig 2 shows the mass loss of an aluminium container immersed in solutions of different pH.

It is clear that acidic or alkaline foods will tend to dissolve more aluminium and it is probable that between pH 3 and pH 8, little of the element will be dissolved. (On the other hand, *small* amounts may be, and even these quantities may be serious.)

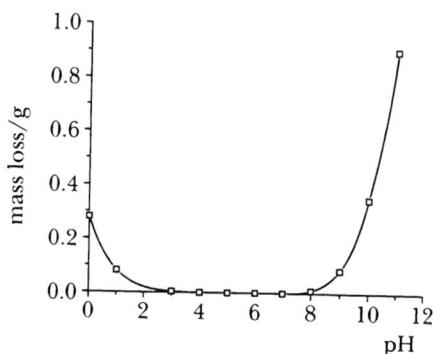

FIG 2

Bioavailability

For any constituent of food to be absorbed by the body, it must be able to pass through the wall of the stomach or the intestine. In other words, it must be able to pass across cell membranes. The cell membranes of mammals are made of non-polar organic molecules, and only non-polar molecules can dissolve in and thus pass through these membranes. That is why no ionic species are soluble in the cell membrane. Those ions that the body needs – sodium, potassium, iron and so on – have a specific absorption mechanism. In each case, there is a particular molecule in the membrane which binds to the ion and carries it across the membrane. There is a special sodium-binding substance, another for potassium, and so on. Aluminium is not required by the body and there is no specific aluminium-binding molecule, so it should not be absorbed.

Unfortunately, an aluminium ion is sufficiently like an iron (III) ion for it to be mistaken for it. As a result it can bind to the iron transport protein (transferrin) and be absorbed. Presumably this is by interaction of the Fe^{3+} (or Al^{3+}) ions with negatively charged groups on the protein. Transferrin cannot take up insoluble iron compounds. The iron (or aluminium) must be in solution. So if there are any compounds in the gut which can form soluble complexes with the iron (or aluminium), this may help them to be absorbed.

CONTINUED

Sand castles and mud huts © 1991 Jeffrey Hancock, published by Hodder and Stoughton Educational

But aluminium absorption is extremely inefficient or, to put it another way, the body is extremely efficient at excluding it. We ingest 20–30 mg of it each day but excrete *most* of it. How much we retain in our bodies and its precise effect are not yet known.

Questions

6. Why do potatoes cooked in an aluminium saucepan show a much smaller increase in their aluminium content than rhubarb or apples?

7. Aluminium metal will react with both an acid and a base.

(a) Write equations for each of these reactions.

(b) Use them to explain the shape of the graph in fig 2.

8. Aqueous aluminium ions complex strongly with fluoride ions (which are added to water to help prevent tooth decay).

(a) What is the formula of the complex?

(b) Write an equilibrium to show its formation.

(c) Why do F^- ions complex more strongly than water with Al^{3+} ions?

(d) Fluoride ions increase the tendency of aluminium metal to dissolve in water. Explain why.

9. Phosphate ions (in food additives numbers E338–341) also complex strongly with aluminium ions, forming species such as $AlPO_4$, $[Al(PO_4)_2]^{3-}$ and so on.

(a) Suggest the formula of another aluminium-phosphate complex.

(b) $AlPO_4$ is insoluble in water, while $[Al(PO_4)_2]^{3-}$ is soluble. Explain this.

(c) Why do phosphate ions complex so strongly with aluminium ions?

(d) Phosphoric acid is added to colas. Suggest *two* reasons why this might be inadvisable if the cola is to be sold in an aluminium can.

10. Iron and aluminium both bind to the iron transport protein, transferrin. Some data relating to this are given in table 2.

TABLE 2

	Al^{3+}	Fe^{3+}	Ga^{3+}
Electron configuration	[Ne]	[Ar]$3d^5$	[Ar]$3d^{10}$
E°/volts, M^{3+} (aq), M (s)	−1.66	−0.04	−0.53
Electronegativity	1.5	1.8	1.6
Ionic radius/nm	0.057	0.067	0.062
Covalent radius/nm	0.125	0.117	0.125
Bond enthalpy/kJ mol^{-1}			
M−O	585	390	363
M−Cl	498	293	485
M−H	285		288

(a) Why is aluminium mistaken by the body for iron?

(b) Transferrin binds aluminium ions about 10^{10} times more weakly than it binds iron. Why do you suppose that might be so?

(c) Gallium ions are bonded to transferrin almost as strongly as Fe^{3+} ions. Suggest why.

15. Rocket fuels

From the Messerschmidt Me 363 of World War II to the Apollo lunar
lander and the Space Shuttle; hydrazines have fuelled them all. The
reason is not hard to find. Hydrazine is an endothermic compound:

$$N_2(g) + 2H_2(g) \rightarrow N_2H_4(l) \quad \Delta H_f^\circ = +50.6 \text{ kJ mol}^{-1} \qquad [1]$$

so the decomposition of hydrazine back to its elements releases much
energy. In addition, hydrazine is readily oxidised.

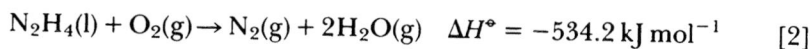

$$N_2H_4(l) + O_2(g) \rightarrow N_2(g) + 2H_2O(g) \quad \Delta H^\circ = -534.2 \text{ kJ mol}^{-1} \qquad [2]$$

This instability and exothermic oxidation make production difficult,
but the USA, Europe and Japan need 15–20 000 tonnes a year. The
manufacturing process involves oxidation of ammonia with sodium
chlorate (I) solution and it proceeds in two steps. The first is the same as
the reaction that occurs if a chlorine bleach and an ammonia-based
household cleaner are mixed:

$$NH_3 + NaOCl \rightarrow NaOH + NH_2Cl \quad \text{(fast)} \qquad [3]$$

(Don't try it! Chloramine fumes are unpleasantly choking.) This is
followed by:

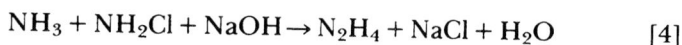

$$NH_3 + NH_2Cl + NaOH \rightarrow N_2H_4 + NaCl + H_2O \qquad [4]$$

There is one serious side reaction, too, which destroys the hydrazine
produced:

$$N_2H_4 + 2NH_2Cl \rightarrow 2NH_4Cl + N_2 \qquad [5]$$

This reaction [5] is catalysed by minute traces of Cu^{2+} ions.
Fortunately, addition of a protein such as gelatine mops up these Cu^{2+}
ions, thus preventing reaction [5] from occurring. A solution of
hydrazine of about 0.5% concentration by mass is produced;
distillation eventually gives pure hydrazine hydrate, $N_2H_4 . H_2O$.
Hydrazine itself can be obtained by mild dehydration.

Rocket motors have been built in which hydrazine reacts with oxygen
(reaction [2]). Only hydrogen gives more thrust per kilogram of fuel.
But oxygen gas takes up too much space, and liquid oxygen boils at
−183°C. Dinitrogen tetraoxide (N_2O_4) boils at 21°C, and so is easier to
handle.

$$2N_2H_4(l) + N_2O_4(g) \rightarrow 3N_2(g) + 4H_2O(g) \qquad [6]$$

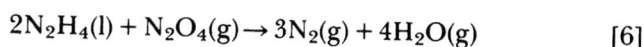

Reaction [6] is also exothermic (see question 7) so can be used for
propulsion. But hydrazine is not the ideal fuel. It melts at 2°C, so if the
temperature drops below this, the fuel will solidify. For this reason,
mixtures with methylhydrazine (CH_3NHNH_2) or the unsymmetrical
dimethylhydrazine (($CH_3)_2NNH_2$) are often used, as these have lower
melting points. The Apollo moon lander used three tonnes of
methylhydrazine for the actual moon landing.

Best of all, hydrazine and its derivatives can be used alone, without an
oxidiser. The decomposition of gaseous hydrazine is less exothermic
than its combustion or reaction with N_2O_4, but it is quite adequate for
small manoeuvres in space – such as an alteration in the pitch of the
spacecraft – so only one fuel tank is needed.

But rocket fuels are not the only use for hydrazines. Hydrazine itself
is added to boiler feed water, and reaction [2] occurs, thus removing
virtually all traces of the highly corrosive dissolved oxygen. The

CONTINUED

antibiotic isoniazid, which has all but eradicated tuberculosis, is made from it (fig 1).

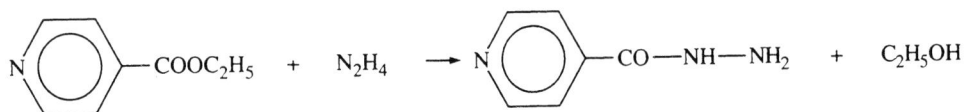

FIG 1

Questions

On nitrogen chemistry and energetics

1. Explain *briefly*:

    (a) what is meant by 'an endothermic compound',

    (b) how pure hydrazine hydrate (b.p. 118°C) is obtained from the reaction mixture after reaction [4].

2. What is the oxidation number of nitrogen in:

    (a) NH_3?   (b) N_2H_4?   (c) NH_2Cl?   (d) N_2?
    (e) NH_4Cl?   (f) N_2O_4?

3. (a) If liquid hydrazine decomposes to its elements, energy is released. How much?

    (b) If decomposition like this is so favourable, why doesn't hydrazine decompose spontaneously?

    (c) Suggest one way in which you could bring about this decomposition. Explain your answer.

    (d) The engines that spacecraft use for manoeuvring operate by passing hydrazine gas over a cold metal gauze, and this causes the decomposition to occur spontaneously. Explain what happens.

4. In many ways, hydrazine is similar to ammonia.

    (a) Draw the structure of ammonia, showing all the valence electrons and labelling the bond angles.

    (b) Do the same for hydrazine.

    (c) In water, ammonia is a base. Write an equilibrium to illustrate this.

    (d) What feature of the NH_3 molecule enables it to act like this?

    (e) Would you expect hydrazine also to act as a base in water? Explain, with an equilibrium.

5. Explain the following:

    (a) N_2H_4 has a higher boiling point (386K) than C_2H_4 (169K).

    (b) N_2H_4 is soluble in water, C_2H_4 is not.

    (c) Methylhydrazine (CH_3NHNH_2) has a lower melting point than hydrazine. Unsymmetrical dimethylhydrazine (($CH_3)_2NNH_2$) has a lower melting point still.

6. You are given the following bond energies: $C-C$ 347; $N-N$ 158; $O-O$ 144; $F-F$ 158; $N-H$ 391; $H-H$ 436; $C\equiv C$ 838; $N\equiv N$ 945 kJ mol^{-1}.

    (a) The $N\equiv N$ bond is stronger than the $C\equiv C$ bond. Why?

    (b) It has been suggested that the weakness of the $N-N$, $O-O$ and

CONTINUED

F–F bonds compared to the C–C bond is a result of repulsions between non-bonded electron pairs. Explain this.

(c) Use the bond energies to calculate ΔH for the reaction:

$$N_2H_4(g) \rightarrow N_2(g) + 2H_2(g)$$

(d) What particular feature(s) of nitrogen chemistry cause this reaction to release energy?

7. The standard enthalpy of formation of dinitrogen tetraoxide is $+9.2 \text{ kJ mol}^{-1}$, and of gaseous water, $-241.8 \text{ kJ mol}^{-1}$.

(a) Calculate the enthalpy change, per mole of hydrazine, for the reaction of hydrazine and dinitrogen tetraoxide (reaction [6]).

(b) Compare your answer with the enthalpy change when hydrazine is oxidised with oxygen (reaction [2]), and comment.

8.* Ammonia reacts with esters or acid chlorides to form amides:

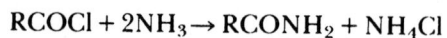

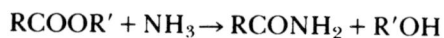

$$RCOCl + 2NH_3 \rightarrow RCONH_2 + NH_4Cl$$

$$RCOOR' + NH_3 \rightarrow RCONH_2 + R'OH$$

(a) The carbonyl group (the CO group) is polar. Explain what this means.

(b) Draw the mechanism for the first step of either reaction above.

(c) Suggest a mechanism for the first step in synthesis of isoniazid.

Sand castles and mud huts © *1991 Jeffrey Hancock, published by Hodder and Stoughton Educational*

16. Most dangerous, most vital

Over a million tonnes of hydrogen fluoride (HF) are made each year throughout the world. The starting material is fluorspar, which is treated with concentrated sulphuric acid:

$$CaF_2 + H_2SO_4 \rightarrow 2HF + CaSO_4 \qquad [1]$$

The reaction is done in a horizontal steel drum about 30 metres long, rotating at about one revolution per minute. The gas is condensed, then redistilled to produce 99.9% pure HF.

The dangers of HF stem from the very properties that make it so useful. Although the solid and liquid consist of zigzag chains held together by hydrogen bonds, in the liquid the average chain is no more than 3.5 molecules long and there are also $(HF)_6$ rings. These relatively small units mean that the liquid is less viscous than H_2SO_4 or even H_2O. HF forms several strongly bonded hydrates, such as $HF \cdot H_2O$. These cause liquid HF to be a strong dehydrating agent. In solution in water, HF is a weak acid (unlike the other hydrogen halides), because the $HF \cdot H_2O$ dissociates only to a small extent:

$$H_2O + HF \rightleftharpoons [H_2O \cdot HF] \rightleftharpoons [H_3O^+F^-] \rightleftharpoons H_3O^+(aq) + F^-(aq) \qquad [2]$$

with the final equilibrium lying well to the left. However, as the concentration of the acid increases, another equilibrium is set up:

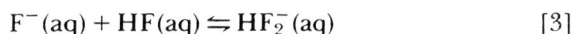

$$F^-(aq) + HF(aq) \rightleftharpoons HF_2^-(aq) \qquad [3]$$

It lies to the right, and pulls the other equilibria over, thus increasing the concentration of oxonium ions, H_3O^+.

Virtually any contact with the acid is damaging. It is particularly nasty in that the characteristic white burns and excruciating pain develop only after some hours. This is caused by a combination of the easy penetration of the skin, dehydrating effect and low pH. The extreme pain and very slow healing are caused by the removal of calcium ions from the skin tissues by the formation of insoluble calcium fluoride (CaF_2). First aid must be given immediately: thorough washing under the tap for 15 minutes, followed by dressings containing magnesium compounds. In extreme cases, injections of calcium gluconate under and around the burn are required.

Faced with all this, why on earth is HF made in such quantities? It is the first step in the manufacture of fluorine which can only be made by electrolysis of HF/KF mixtures. Fluorine and hydrogen fluoride are used for so many vital processes. Consider the following.

Aluminium is made by electrolysis of aluminium oxide (Al_2O_3) in molten cryolite (Na_3AlF_6). Although the cryolite is in theory only a solvent, inevitably some decomposes and must be replaced. Without fluorine, no aluminium.

Other metal fluorides, such as sodium fluoride, may be added to your drinking water supply. Tin (II) fluoride is a toothpaste ingredient to the extent of 0.4%. Between them, these fluorides have reduced tooth enamel decay dramatically, with no ill effects.

Think of tetrafluoroethene ($CF_2=CF_2$) which is polymerised to make PTFE, a uniquely inert polymer (see Section 27: Spare parts, p. 101).

And remember anaesthetics, almost all of which are now organic fluorine compounds (see Section 6: Anaesthetics, p. 26). Appendicitis may be painful; but think how painful it would be without an anaesthetic!

CONTINUED

Questions

On fluorine chemistry

1. Solid HF contains chains of the molecules (fig 1):

    (a) Explain why these chains form.

    (b) Why aren't the chains linear? In other words, why isn't the angle α 180°?

    (c) Suggest a value for α.

FIG 1

    (d) Draw one of the $(HF)_6$ rings found in liquid hydrogen fluoride.

2. (a) Draw a diagram of the hydrated F^- ion.

    (b) The F^- ion also forms the HF_2^- ion. Draw a diagram to show the bonding in this.

    (c) Equilibrium [3] shows that the formation of the HF_2^- ion is more favourable than the formation of F^-(aq). Why is this?

3. (a) Why does HF form such a stable hydrate?

    (b) In fact, it also forms $H_2O . HF$ and $H_2O . 4HF$. Draw the bonding in $H_2O . 4HF$ and suggest why $H_2O . 5HF$ doesn't form.

4. (a) The extreme pain and slow healing of fluoride burns are caused by the removal of Ca^{2+} ions from the skin tissues to form CaF_2. This is a solid in the body because it has a very stable lattice. Why do fluorides have such stable lattices, more stable than chlorides, for example?

    (b) Emergency treatment for fluoride burns uses magnesium or calcium ions, either washed on or injected. Suggest how this works.

    (c) Teeth are composed of hydroxyapatite, $Ca_5(PO_4)_3(OH)$. One part per million of fluoride ions in the water supply (which often occur naturally, but in some places are added) reduces tooth decay by replacing some of the OH^- ions with F^- ions. This tends to make the tooth material harder and less soluble in water (or saliva). Can you suggest why this might be so?

5. Cryolite contains the $[AlF_6]^{3-}$ ion.

    (a) Draw a diagram to show the arrangement of the bonding electrons in this complex ion.

    (b) How many electrons are there in the bonding orbitals of the aluminium atom in the complex? How is this possible?

    (c) What shape will it be?

    (d) Chlorine forms $[AlCl_4]^-$, but $[AlCl_6]^{3-}$ does not exist. Why not?

    (e) Boron forms only $[BF_4]^-$. $[BF_6]^{3-}$ does not exist. Why not?

6. (a) Tin(II) fluoride is made from the metal or the oxide. Write equations for these reactions.

    (b) If fluorine is reacted with tin metal, the product is tin(IV) fluoride. Suggest why, and write an equation for its formation.

7. Both fluorine and aluminium are made by electrolysis. Why aren't chemical production methods used which would probably be cheaper?

8. Why is polytetrafluoroethene so inert?

17. Life's little ironies

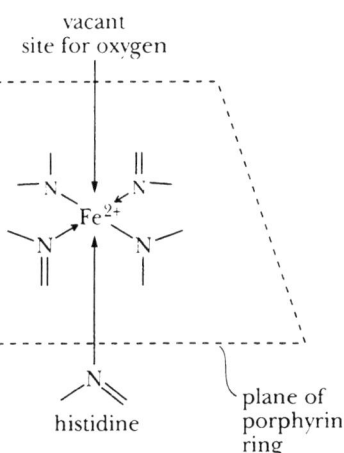

FIG 1

The human body contains about four grammes of iron; that is, about 0.006% by mass, but this tiny amount is essential. For example, the hydrogen cyanide used in the gas chambers of some American states works by attacking just a part of it. One of the harmful effects of cigarette smoking is due to a reaction with another portion of it.

The best-known coloured compound of iron in the body is haemoglobin. It is responsible for the colour of red blood cells, and absorbs oxygen in the lungs and carries it to the muscles. Each molecule consists of four units, each containing an Fe^{2+} ion, forming a complex with a protein and a porphyrin group. The porphyrin is a flat molecule with four nitrogen atoms in the right positions to co-ordinate with four of the six co-ordination sites of the Fe^{2+} ion. The fifth position is occupied by a nitrogen atom of the amino acid histidine, part of the protein (fig 1). The sixth position is vacant (although usually occupied by weakly bonded water), and available to bind oxygen. Thus the oxygen is absorbed by the haemoglobin, and can be carried by the blood to the muscles. There it is given up to myoglobin (another iron-containing protein, somewhat similar to haemoglobin) which stores it in the muscles, ready for use.

Interestingly, although oxygen is normally capable of oxidising Fe^{2+} to Fe^{3+}, it can't oxidise the Fe^{2+} bound to haemoglobin. Nitrite ions and hydrogen peroxide can cause this oxidation, however, producing methaemoglobin, in which the iron exists as the Fe^{3+} ion. Methaemoglobin cannot bind and transport oxygen, so if too much of it is formed in the body, oxygen cannot get to the muscles or the brain. This explains one problem of excessive use of fertilisers, which can give rise to nitrites in the water supply. Babies are particularly vulnerable to this: in some areas of the country, bottled water is supplied for babies if nitrite levels are too high.

Haemoglobin can be readily poisoned. Carbon monoxide (CO) acts by complexing with the Fe^{2+} in the same way as oxygen, but more strongly.

Questions

On transition element chemistry

1. Transition elements differ from the other elements in the periodic table in several characteristic ways. Find an example in the passage of each of these:

    (a) variable oxidation states,
    (b) formation of complexes,
    (c) catalytic activity,
    (d) coloured compounds.

2. Give the electronic configuration of:

    (a) the Fe atom,
    (b) the Fe^{2+} ion,
    (c) the Fe^{3+} ion.

CONTINUED

3. (a) What is a ligand?

(b) Ligands which form complexes with Fe^{2+} or Fe^{3+} ions include H_2O, CN^- and PF_3. What do the electronic configurations of these have in common?

(c) How do these ligands bind to the ions?

(d) On 10 August 1990 six men were overcome by methane in a sewer beneath Edinburgh, and one died. Methane kills by suffocating and not by complexing with metal ions as CO and CN^- ions do. Why can't the methane complex in this way?

4. (a) When dissolved in water, Fe^{2+} and Fe^{3+} ions form hydrated ions. What does this mean? Give possible formulae for the hydrated ions.

(b) These hydrated ions undergo hydrolysis. What does this mean? Give suitable equilibria. Which hydrated ion undergoes the most extensive hydrolysis? Why?

(c) The stomach is acidic (pH <2), the intestine slightly alkaline. How will this affect the extent of hydrolysis of the two ions?

Energy for all the processes of a mammal's body comes from the oxidation of food. This occurs in several steps, with intermediate substances being oxidised or reduced on the way. So, for example, the first step of oxidation of compound R might involve the reduction of compound S:

$$R + S \rightarrow \text{oxidised R} + \text{reduced S}$$

Reduced S might then be reoxidised by reaction with T:

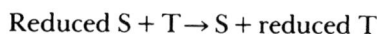

$$\text{Reduced S} + T \rightarrow S + \text{reduced T}$$

The reduced T is then reoxidised, and so on, involving several steps, until in the last step:

$$\text{Reduced X} + O_2 \rightarrow \text{water} + X$$

These processes occur in structures inside cells called mitochondria. The last three compounds of this series are called cytochromes, and all of them contain Fe^{2+}, converted to Fe^{3+} on oxidation and back to Fe^{2+} again on reduction.

These cytochromes are the last steps in oxidation of foodstuffs, so if they are damaged, energy production rapidly stops. Cyanide ions are fatal because they damage the cytochromes by interacting with the Fe^{2+} or Fe^{3+} ions in them.

Iron also plays a key role in several enzymes. Perhaps catalase is the most well known. It occur in blood and catalyses the breakdown of hydrogen peroxide:

$$2H_2O_2 \rightarrow 2H_2O + O_2$$

Exactly why this is valuable to the body is not clear. One possibility lies in the fact that hydrogen peroxide oxidises the Fe^{2+} in haemoglobin to Fe^{3+}, producing methaemoglobin. Since methaemoglobin cannot bind oxygen, the presence of hydrogen peroxide – which can be formed by some bacteria – could prevent oxygen transport. Catalase prevents this, by breaking down any H_2O_2 as it is produced. Some individuals, especially in Korea and Japan, have low catalase levels, or in a few cases, none of it at all. It seems to cause them few problems, except perhaps in those regions of the body where bacteria are common, around the teeth. These people with low (or zero) catalase levels tend to suffer from ulceration of the gums.

Since iron is so important for body processes, it is vital to make sure

CONTINUED

Sand castles and mud huts © 1991 Jeffrey Hancock, published by Hodder and Stoughton Educational

that there is an adequate stock of it in the body. Fortunately, almost all of it is recycled. For example, when red blood cells are broken down, new haemoglobin is synthesised using the iron from the old cells. Iron is lost in very small amounts in urine or by bleeding, normally less than one milligram a day. Menstruation is more significant, and may cause a daily loss of three or four milligrams. Young girls – who need iron for growth but who are also losing it in their periods – and pregnant women have particularly high daily requirements. It is not surprising that perhaps 10% of women are anaemic, and 30% more on the verge of it.

Many foods are rich in iron, but it cannot be absorbed unless it is in solution in the gut contents. In the stomach, of pH 1–2, free hydrated Fe^{2+} and Fe^{3+} ions can exist in solution, but once the food has passed into the duodenum (pH 8 or so), the hydroxides $Fe(OH)_2$ and $Fe(OH)_3$ will be formed. These are much less soluble, so they are less available for absorption. Studies with radioactive iron show that many other food constituents help iron absorption from the duodenum. Vitamin C reduces $Fe(OH)_3$ to the more soluble $Fe(OH)_2$, while some sugars, citric acid or amino acids form soluble complexes, preventing the iron from precipitating out in alkaline solution, and thus keeping it available for absorption.

Questions

5. Two symptoms of anaemia are paleness and tiredness. Give explanations for these based on the information in the passage.

6. (a) Carbon monoxide and cyanides are fatal, but whereas you have to breathe air containing 1500 ppm of CO for an hour for it to kill you, HCN at only 100 ppm is fatal after an hour. Why is cyanide so much more toxic? (In other words, why do CN^- ions bind so much more strongly than CO molecules to iron ions?)

(b) Cyanide ions actually complex more strongly with Fe^{3+} ions than with Fe^{2+} ions. Why do you think this is so?

(c) Quite small amounts of cyanide ions can kill you by complexing with the Fe^{3+} ion in the last of the cytochromes. If the cyanide has been swallowed, death is fairly slow. A good antidote is injections of solutions of cobalt(II) salts. How do they work?

(d) Inhaled HCN is rapidly absorbed and is fatal in minutes. Inhaled amyl nitrite is the best emergency antidote. Explain how it works. (*Hint*: The nitrite oxidises some of the body's haemoglobin to methaemoglobin, containing Fe^{3+}. There is much more iron in the haemoglobin than in the cytochromes.)

7. (a) Some long distance runners have been accused of 'blood doping', in which 0.5–1 dm^3 of their blood is removed several weeks before a big race and stored. The red cells are then concentrated and put back a day or two before the race. How might this help their performance?

(b) Why would it be better to run a marathon in the country rather than in the city?

(c) Could a catalase-deficient person make a good athlete? Explain.

8. (a) Look at table 1. Is it easier to oxidise $[Fe(CN)_6]^{4-}$ to $[Fe(CN)_6]^{3-}$, or $[Fe(H_2O)_6]^{2+}$ to $[Fe(H_2O)_6]^{3+}$?

(b) Oxygen cannot oxidise haemoglobin to methaemoglobin (that is $Hb-Fe^{2+}$ to $Hb-Fe^{3+}$), but hydrogen peroxide can. Suggest a value for the standard electrode potential of the methaemoglobin, haemoglobin couple; that is, for the couple $[Hb-Fe^{3+}]$, $[Hb-Fe^{2+}]$.

TABLE 1

Standard electrode potentials	$E°$/volts
$[Fe(CN)_6]^{3-}$, $[Fe(CN)_6]^{4-}$	$+0.34$
$[Fe(H_2O)_6]^{3+}$, $[Fe(H_2O)_6]^{2+}$	$+0.77$
$[\frac{1}{2}O_2 + 2H^+]$, H_2O	$+1.23$
$[H_2O_2 + 2H^+]$, $2H_2O$	$+1.77$

CONTINUED

9. The last three steps of the oxidation of food in cell mitochondria involve cytochromes a, b and c, whose $E°$ values are given in table 2.

In what *order* do these cytochromes take part in the oxidation sequence?

10.* (a) Oxygen bonds to the Fe^{2+} ion in haemoglobin:

$$O=O \rightarrow Fe^{2+}$$

It is not yet certain exactly what shape this is. What shape would you expect? What would you expect the $O-O-Fe$ bond angle to be? (Use the normal rules for shapes of molecules.)

   (b) Presumably CO bonds to haemoglobin in the same way:

$$OC \rightarrow Fe^{2+}$$

What shape will this be? Suggest a value for the $O-C-Fe$ bond angle.

TABLE 2

Standard electrode potentials	$E°$/volts
Cytochrome a Fe^{3+}, Fe^{2+}	+0.29
Cytochrome b Fe^{3+}, Fe^{2+}	+0.04
Cytochrome c Fe^{3+}, Fe^{2+}	+0.26

Sand castles and mud huts © 1991 Jeffrey Hancock, published by Hodder and Stoughton Educational

18. Ozone problems

The five essential chemical reactions that occur in the stratosphere – a layer of the atmosphere between about 10 and 50 kilometres above the Earth's surface – were first proposed by the physicist, Sydney Chapman, in 1930.

Oxygen molecules absorb ultraviolet light and split to form oxygen atoms:

$$O_2 + light \rightarrow \cdot O \cdot + \cdot O \cdot \qquad [1]$$

These oxygen atoms might recombine (reversing reaction [1]), or one oxygen atom may combine with another oxygen molecule, to form a molecule of trioxygen (ozone):

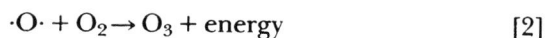

$$\cdot O \cdot + O_2 \rightarrow O_3 + energy \qquad [2]$$

The trioxygen can now undergo two reactions, either with an oxygen atom:

$$\cdot O \cdot + O_3 \rightarrow O_2 + O_2 \qquad [3]$$

or it can split up in a reversal of reaction [2]; the energy required is supplied by UV light again:

$$O_3 + light \rightarrow \cdot O \cdot + O_2 \qquad [4]$$

Reaction [4] explains the vital role of trioxygen in the upper atmosphere; because it absorbs ultraviolet below about 320 nm, it reduces the amount of UV that reaches the Earth's surface (fig 1).

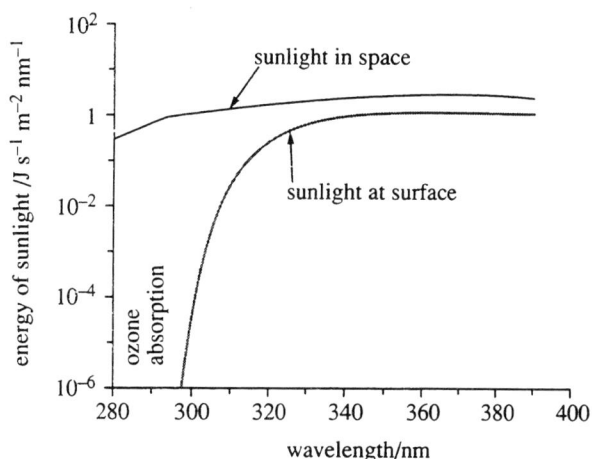

FIG 1

Chlorofluorocarbons (CFCs) have widespread uses, most notably as aerosol propellants, blowing agents for plastic foams and refrigerant liquids. The stability that makes them so safe to use is the cause of the problems when they escape into the atmosphere. Nothing happens to them in the lower layers of the atmosphere (the troposphere). They don't dissolve in rain and are not broken down by sunlight. So they just remain in the atmosphere, slowly spreading out and upwards.

Once the CFCs reach the stratosphere, several reactions start to happen. First, they are photolysed, that is, they are broken down by light. Taking CCl_3F as an example:

$$CCl_3F + light \rightarrow Cl \cdot + \cdot CCl_2F \qquad [5]$$

CONTINUED

The chlorine atom can now react with a trioxygen molecule:

$$Cl\cdot + O_3 \rightarrow ClO\cdot + O_2 \qquad [6]$$

and the $ClO\cdot$ produced can combine with an oxygen atom:

$$ClO\cdot + \cdot O\cdot \rightarrow Cl\cdot + O_2 \qquad [7]$$

These reactions are doubly troublesome. Reaction [6] decomposes ozone, and [7] mops up $\cdot O\cdot$ that might have reacted with O_2 to make more ozone by reaction [2].

If the ozone is destroyed, much of the UV that it now filters out will reach the ground. What effect will this have? It will cause eye cataracts and skin cancers in humans, but more serious, perhaps, is its effect on plants. Although this has not been studied much yet, early work shows that crop yields may be reduced, trees stunted and plankton damaged.

There is much evidence that the ozone layer is being depleted. Holes in the ozone layer have been seen to open up every spring over Antarctica. In response, the 'Montreal Protocol' has been ratified by 80 countries, requiring them to phase out CFC use by the year 2000. This isn't the end of the problem as the lifetimes of CFCs already in the atmosphere are between 70 and 380 years, so the problem will be with us for a long time. A short-term solution is to use chlorofluorohydrocarbons, such as $CHClF_2$ and $C_2H_3ClF_2$, which are broken down in the troposphere. They may cause other atmospheric damage so completely safe replacements will have to be found.

Questions

On reactions of free radicals

1. The reactions of this passage all involve free radicals, such as $\cdot OH$ or $Cl\cdot$.

   (a) What is a free radical?

   (b) Give two characteristics of free radicals.

   (c) In equation [1] of the Chapman cycle, why is the oxygen atom written as $\cdot O\cdot$? What does this tell you about the chemistry of the $\cdot O\cdot$?

2. (a) The frequency of light required to break a bond of energy E is given by the Planck equation: $E = L \times h \times f$

where E = the bond energy in joules mol^{-1}
         L = Avogadro's constant ($= 6.02 \times 10^{23}\,mol^{-1}$)
         h = Planck's constant ($= 6.63 \times 10^{-34}\,JHz^{-1}$)
         f = the frequency of the light.

Calculate the frequency of the light required to break the oxygen bond, given that its energy is $497\,kJ\,mol^{-1}$, and

   (b) the wavelength of the light required (given that $c = f \times \lambda$, where λ = the wavelength of the light, and c = the speed of light = $3.00 \times 10^8\,m\,s^{-1}$).

   (c) As you go up from the surface of the Earth, the temperature falls, but in the stratosphere, it rises again. Use reactions [1–4] to suggest an explanation for this rise.

3. (a) How much energy will be released (per mole) when two $\cdot O\cdot$ atoms recombine?

   (b) How much energy is required to split the O_2 molecules back into $\cdot O\cdot$ atoms?

   (c) In practice, the recombination $\cdot O\cdot + \cdot O\cdot \rightarrow O_2$ requires the presence of a third molecule to absorb the energy released. Explain why.

CONTINUED

Sand castles and mud huts © 1991 Jeffrey Hancock, published by Hodder and Stoughton Educational

4. (a) The reaction of CH_4 and Cl_2 in the presence of light is the classic example of a free radical chain reaction. Show how the reaction proceeds.

(b) What is meant when it is described as a *chain* reaction?

(c) Find an example of a chain reaction in the above passage, making clear why it is a chain.

(d) What stops or slows down a chain reaction? Why doesn't this happen much for the chain reaction you have written in (c)?

5. (a) When spread on the land, nitrogen fertilisers are lost in two main ways. They are leached by the rain into rivers, lakes and so on, and denitrifying bacteria convert them into gaseous oxides of nitrogen. These gaseous nitrogen oxides rise up through the atmosphere, and when they reach the stratosphere, undergo further reactions, to produce nitrogen monoxide (NO). The nitrogen monoxide can react with trioxygen in a similar way to chlorine atoms. Construct similar equations to [6] and [7], but involving NO instead of chlorine atoms.

(b) Draw the electron arrangement in nitrogen monoxide. What feature is common to both chlorine atoms and NO?

6. The increased UV light reaching the Earth's surface as the ozone layer is depleted will cause increased skin cancers and cataracts, and will probably inhibit plant growth. Which do you think we should be more concerned about? Why?

Ozone in the stratosphere makes life possible, but at sea level it is vile. It became famous as a component of the smog of Los Angeles, where it reaches such high concentrations that it can be smelt in the city's air. It is a powerful oxidising agent, attacking any organic material, from lungs to leather and from rubber to trees. Research is beginning to explain how it is formed, along with a cocktail of other pollutants, by the action of sunlight on nitrogen oxides and unburnt hydrocarbons from car exhausts. For example:

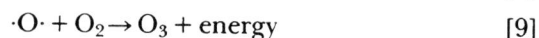

$$NO_2 + light \rightarrow \cdot O \cdot + NO \qquad [8]$$

$$\cdot O \cdot + O_2 \rightarrow O_3 + energy \qquad [9]$$

A series of reactions now ensues, converting hydrocarbons into aldehydes and generating $\cdot OH$ radicals, which then react further:

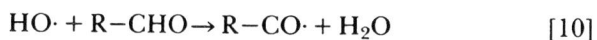

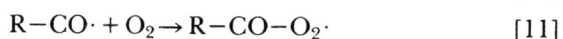

$$HO \cdot + R-CHO \rightarrow R-CO \cdot + H_2O \qquad [10]$$

$$R-CO \cdot + O_2 \rightarrow R-CO-O_2 \cdot \qquad [11]$$

$$R-CO-O_2 \cdot + NO_2 \rightarrow R-CO-O-O-NO_2 \qquad [12]$$

If the $R-$ in this last compound is CH_3-, the product is peroxyethanoyl nitrate (also called peroxyacetyl nitrate, or PAN). It is a particularly horrid chemical causing streaming eyes and attacking vegetation.

British cities lack both the large numbers of cars and the bright sunlight, or they did until the long hot summers of 1989 and 1990. While not yet matching the Los Angeles figures, summer ozone levels are now a cause for concern.

Let's be clear about this: it's not chemistry that has created the problem, it is the car. There are only two solutions: abandoning the car or using some very sophisticated chemistry to solve the problem.

CONTINUED

Questions

7. The levels of NO_2 and ozone in Los Angeles' air for one summer's day are plotted in fig 2.

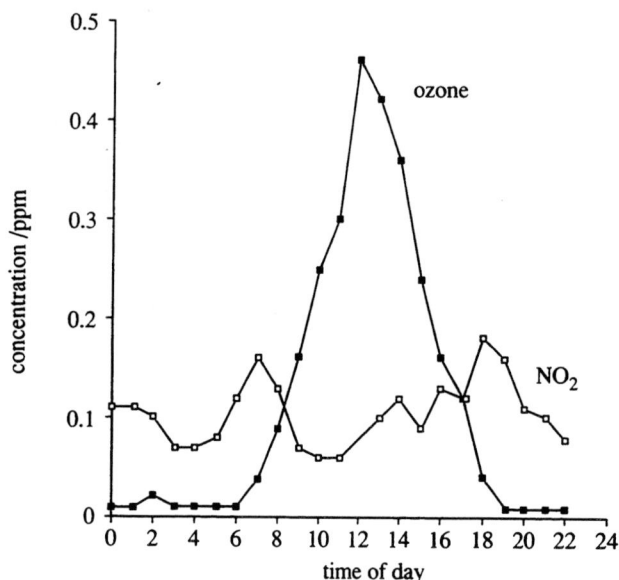

FIG 2

Use the information in the passage to suggest why:

(a) NO_2 levels are so high between 5 a.m. and 9 a.m., and 3 p.m. and 9 p.m.

(b) ozone levels start to rise at 6 a.m.

(c) ozone levels fall off after midday. (Surely ozone will go on being produced all day?)

8. Ozone levels in British cities are monitored regularly. One method of doing this is by light absorption. A beam of light is passed through the air, and the amount of light absorbed at a characteristic wavelength enables the concentration of the gas to be calculated. The instrument in Coventry uses a beam of light 425 metres long, 40 metres above the High Street.

(a) Why is such a long light beam used?

(b) Suggest a suitable wavelength to use to estimate the ozone concentration.

(c) Some figures for the maximum ozone levels in Coventry for August 1990 are given in table 1.

TABLE 1

Date	1	2	3	4	5	6	7
Max ozone level/$\mu g\, m^{-3}$	83	143	150	80	60	44	49

Suggest a reason for the variations.

(d) The Coventry data is expressed as μg of O_3 per m^3; that for Los Angeles is in parts per million by volume. For O_3, 1 ppm $\simeq 2 \times 10^3\ \mu g\ m^{-3}$. How do the maximum values for the two cities compare?

(e) One reason for the difference in values is geographical; Los Angeles is surrounded by hills which trap the city's air. Give two other reasons for the differences.

Sand castles and mud huts © 1991 Jeffrey Hancock, published by Hodder and Stoughton Educational

19. *Fatty foods*

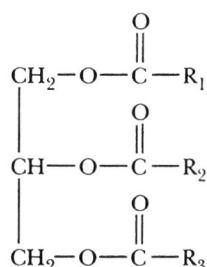

FIG 1

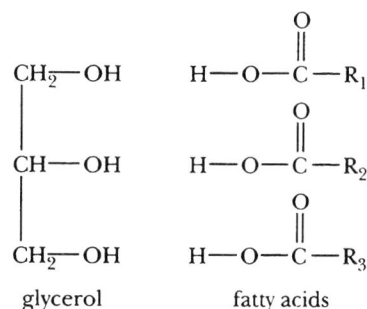

glycerol   fatty acids

FIG 2

Traditional cooking fats were mostly from animal sources: for example, butter (made by churning cream until the fat was deposited), lard (pig fat) and suet (the hard fat packed round the internal organs of sheep). Only olive oil, widely used in Mediterranean countries, was of vegetable origin.

In 1869, Napoleon III announced a competition to find a cheap fat with good keeping qualities – there were no fridges in those days – and the era of artificial fats was born. (The winning margarine was a concoction of milk, beef fat and chopped cow's udders.) Over the next century many new fats were produced, mostly from vegetable sources.

Chemically there is little difference between a fat and an oil, except that solids are called fats, while liquids are usually referred to as oils. They all contain glyceryl triesters. (Fig 1 shows their general structure.) They are broken down in the body to the parent compounds (fig 2).

Compounds of formula $R-COOH$, the fatty acids, determine the nature and effect of the fat or oil. R denotes a hydrocarbon group. It may be saturated, for example, like the C_3H_7- and $C_{17}H_{35}-$ groups. Butyric acid (C_3H_7COOH), more correctly called butanoic acid, makes up 9% of the fat of cow's milk, and stearic acid ($C_{17}H_{35}COOH$) is a major constituent of chocolate. The $R-$ group may be more or less unsaturated. Oleic acid ($C_{17}H_{33}COOH$) has one double bond between the ninth and tenth carbons of the molecule (counting the COOH as the first). This is a very common acid, making up 79% of olive oil, for example. Or the $R-$ group may be polyunsaturated, with two or more double bonds in the molecule. One of the most unsaturated (and important) fatty acids is arachidonic acid ($C_{19}H_{31}COOH$) which occurs widely, for example, in cod-liver oil.

Questions

On properties of the alkenes

1. The general formula of the carboxylic acids is RCOOH, where R is a hydrocarbon group.

   (a) What is a hydrocarbon group?

   (b) What is meant by 'saturated' and 'unsaturated' in this context?

   (c) What is meant by *poly*unsaturated?

2. (a) What is a homologous series?

   (b) Give the general formula for the homologous series of:

      (i) the alkanes,

      (ii) the alkenes.

   (c) The first four carboxylic acids (also called fatty acids) are: methanoic acid, HCOOH; ethanoic acid, CH_3COOH; propanoic acid, C_2H_5COOH and butanoic acid, C_3H_7COOH. Give the general formula for the homologous series of the carboxylic acids.

   (d) Caprylic acid, more properly called octanoic acid, makes up 12% of coconut oil, but only 2% of cows' milk fat. Draw its structure.

CONTINUED

3. The acid

$$CH_3-CH=CH-CH_2-COOH$$
$$\quad 5 \qquad 4 \quad\;\; 3 \qquad 2 \qquad 1$$

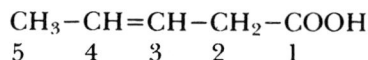

is named 3-pentenoic acid. Draw the structural formulae of two naturally occurring acids:

(a) petroselenic acid ($C_{17}H_{33}COOH$), more properly called 11-octadecenoic acid, which occurs in carrot and celery seeds.

(b) erucic acid (13-docosenoic acid, $C_{21}H_{41}COOH$) which is found in rape seed oil.

4. The saturated acid, stearic acid (octadecanoic acid), has the formula $C_{17}H_{35}COOH$, while linolenic acid is $C_{17}H_{29}COOH$ and linoleic acid, $C_{17}H_{31}COOH$.

(a) How many carbon-carbon double bonds are there in linolenic and linoleic acids?

(b) How could you convert linolenic acid to stearic acid? Give an equation and any special conditions needed.

(c) If you wanted to convert linolenic to linoleic acid, how would you modify your method?

Why do we eat fats? Well, for one thing, because we like them! Look at the fat content of many people's favourite foods. Snacks such as crisps and fast foods like beefburgers are invariably high in fat.

There are good biological reasons for eating fats too. Certainly they are a good source of energy, much better than proteins or carbohydrates. Some fatty acids are vital for body processes. For example, deficiencies in linoleic and arachidonic acids result in skin complaints and weight loss, especially in infants. The role of evening primrose oil in the treatment of eczema may be due to its content of linolenic acid.

In the rich areas of the world at this, the richest, time in the world's history, fat consumption is at record levels. This causes two problems.

If you eat more energy-rich foods than you burn in respiration, the excess is deposited as body fat. Fat in the diet provides more energy than protein or carbohydrate, so a diet rich in fat produces body fat faster than a carbohydrate-rich diet. A slimmer who cuts down his fat intake is likely to lose more weight than one who cuts out bread or potatoes. Unfortunately, there is a lot of 'hidden' fat in food. Everyone knows that butter contains a lot of fat, and so do fried foods. Not so many, perhaps, realise that chocolate, biscuits or meat, even lean meat, also contain fat.

The other problem is coronary heart disease (CHD). The blood supply to the heart is through the coronary arteries, and if these become diseased in any way – usually by a blockage – the blood supply to the heart is interrupted, and the result is a heart attack. Many causes have been suggested for coronary heart disease, such as smoking, high blood pressure, a sedentary life style, or a high intake of dietary fat. Because of this wide variety of possible causes, a major study has collected data from seven different countries. Some results are given in fig 3 on the next page: the incidence of coronary heart disease in men aged 40–59 is plotted against the percentage of saturated fat in their diet. It is clear that the more saturated fat you eat, the more likely you are to die from a heart attack. (The other major – and particularly deadly – risk factor is smoking.)

CONTINUED

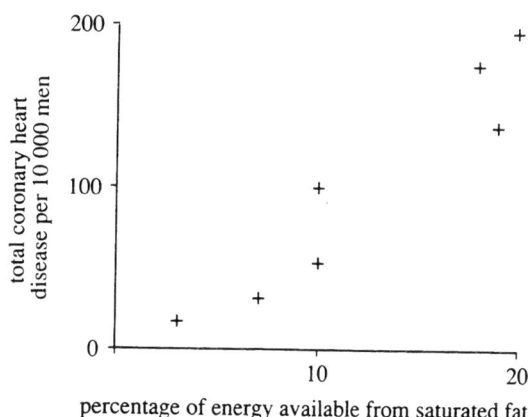

FIG 3

Questions

5. (a) The most abundant fatty acid is oleic acid (9-octadecenoic acid). Draw its structure.

(b) Explain how carbon-carbon single and double bonds are formed. To make their structures clear, draw the molecules of ethane and ethene. Label the bond angles.

(c) Oleic acid is actually *cis*-9-octadecenoic acid. The *trans* isomer, elaidic acid, is much less common. Draw the structure of each molecule, making clear what the difference is between the *cis* and the *trans* forms.

(d) Why can't one form change easily into the other?

6. (a) Construct balanced equations for the complete oxidation to CO_2 and water of:

   (i) the carbohydrate glucose, $C_6H_{12}O_6$

   (ii) the fatty acid, capric (or decanoic) acid, $C_9H_{19}COOH$.

(b) Work out their relative molar masses.

(c) Capric acid produces $6074\,kJ$ when 1 mole of it is burnt, whereas glucose gives only $2816\,kJ\,mol^{-1}$. Calculate the heat produced when one *gram* of each is burnt.

(d) Can you suggest why the fatty acid produces so much more energy than the sugar?

7. A friend of mine had his first heart attack at the age of 38. He immediately gave up smoking, and reduced the amount of saturated fat in his diet. He decided that he would either change from butter to a margarine high in polyunsaturates or drink skimmed milk instead of whole milk. Use table 1 to advise him: which would lower his daily saturated fat intake the most? (He drinks about a pint, or $570\,cm^3$, of milk a day in tea and cereal, and the family of four eats about a pound and a half ($680\,g$) of butter a week.)

TABLE 1

Food	Total fat/%	Saturated fat/%	Polyunsaturated fat/%
Butter	81.0	60.0	2.0
Margarine	80.0	14.0	40.0
Whole milk	3.7	2.6	0.1
Skimmed milk	0.1	0	0

20. Of cholesterol and clots and corpses

Section 19: Fatty foods showed that the more saturated fat you eat, the more likely you are to die from coronary heart disease (CHD).

Why is this? There are several possibilities.

The digestion of fats is a complex affair, and a substantial amount of fat enters the bloodstream as tiny globules. If any fat being carried around in the blood like this is below its melting point, it might form solid deposits in the arteries, deposits which could block the flow of blood. The melting points of the fatty acids vary dramatically with the extent of unsaturation (table 1). The temperature of the human body is about 37°C, so the saturated acids are well below their melting points, while the unsaturated ones are well above theirs. It is clear that the unsaturated acids are not likely to freeze and form solid deposits in the blood, whereas the saturated ones might.

The globules travelling round in the blood after a meal contain various fats and compounds of the free fatty acids with cholesterol. Cholesterol is not a fat. It belongs to the class of compounds called steroids. It is taken in as part of the diet (in meat, eggs and dairy products, for example), and it is also synthesised in the liver, because it has several essential roles in the body's chemistry. For instance, it is vital for the synthesis of steroid hormones, including the sex hormones. It also acts as a sort of detergent, helping to dissolve fats in the blood.

The fatty deposits which form in the arteries of all of us eating a 'Western' diet, and which may even block the artery, contain large amounts of cholesterol. It is not surprising, therefore, that an increased risk of coronary heart disease is strongly linked with increased blood cholesterol levels (fig 1).

TABLE 1   Melting points of fatty acids

	mp/°C
Saturated	
Stearic $C_{17}H_{35}COOH$	70
Arachidic $C_{19}H_{39}COOH$	75
Unsaturated	
Linoleic $C_{17}H_{31}COOH$	−5
Arachidonic $C_{19}H_{31}COOH$	−50

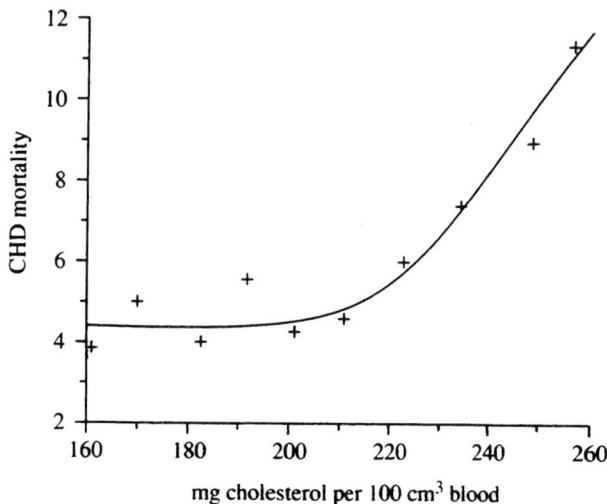

FIG 1

CONTINUED

Sand castles and mud huts © 1991 Jeffrey Hancock, published by Hodder and Stoughton Educational

Why is this? Is it the level of cholesterol or fat intake that is important? Actually, it seems probable that the two are related. A high intake of fat in the diet leads to a high blood cholesterol level, and that, in turn, causes heart disease. If this is right, we should be able to alter the level of blood cholesterol by altering the amount, and the *type*, of fat that we eat. Indeed, many studies have shown this to be so. Increasing saturated fatty acid intake increases blood cholesterol. Interestingly, increasing the intake of *un*saturated fatty acids does not. A government committee has used this fact to make recommendations for a healthier diet (table 2).

TABLE 2

	Present consumption /g per day	Recommendations /g per day
Total fat	104	75–85
Saturated fats	49	25–30
Polyunsaturates	11	15

On average, we should cut our total fat intake by about 25%. (Notice that if we want to reduce our blood cholesterol, we should cut our intake of fat, especially saturated fat. How much cholesterol we eat doesn't seem to matter much.)

Questions

More on alkenes

1. Use the information given in the passage to explain briefly:

    (a) why saturated fat is such a high risk factor for heart disease (you should find *two* possible reasons),

    (b) why a diet low in cholesterol but high in fat does not reduce the risk of heart disease.

2. This question refers to fig 1.

    (a) Does it suggest that cholesterol in the blood is harmful?

    (b) Can you suggest any explanation for the shape of the graph?

3. Instead of writing formulae, chemists often represent molecules by means of stick diagrams. For example:

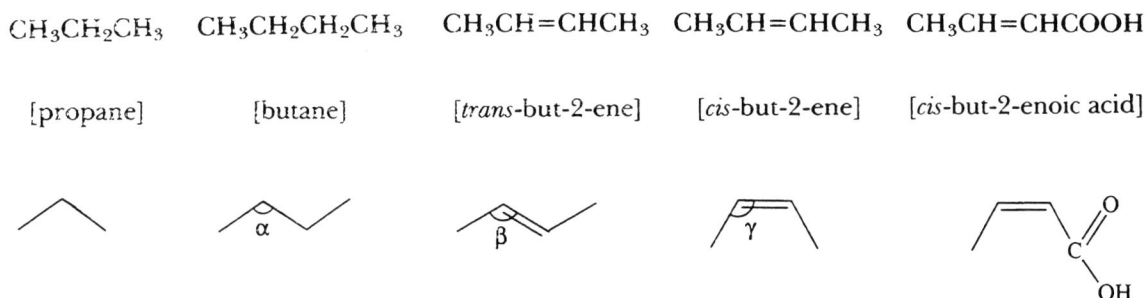

$CH_3CH_2CH_3$ $CH_3CH_2CH_2CH_3$ $CH_3CH=CHCH_3$ $CH_3CH=CHCH_3$ $CH_3CH=CHCOOH$

[propane]   [butane]   [*trans*-but-2-ene]   [*cis*-but-2-ene]   [*cis*-but-2-enoic acid]

FIG 2

    (a) What are the sizes of the angles labelled α, β and γ in the diagrams above?

    (b) Draw stick diagrams of pentane, *cis*-pent-2-ene and *trans*-pent-2-ene.

CONTINUED

(c) When butane is cooled, it condenses and eventually freezes. In the solid, the butane molecules will be regularly packed together, perhaps like those in fig 3. Sketch how the molecules of pentane might pack together in a regular fashion in solid

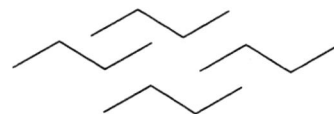

    (i) pentane

    (ii) *cis*-pent-2-ene, and

FIG 3

    (iii) *trans*-pent-2-ene.

(d) The melting points (°C) of these compounds are:

pentane	*cis*-pent-2-ene	*trans*-pent-2-ene
−129	−151	−136

Which has the smallest forces between the molecules? Why do you suppose this is? (Use your sketch diagrams from (b).)

(e) Suggest why *cis* unsaturated acids have much lower melting points than the saturated acid or a *trans* unsaturated acid.

(f) Table 3 gives the percentage of each fatty acid occurring in beef fat and in the liver of the cod.

TABLE 3

	Saturated fatty acids					Mono-, poly-unsaturated				
	C_{14}	C_{15}	C_{16}	C_{17}	C_{18}	C_{14}	C_{16}	C_{18}	C_{20}	C_{22}
Cow	4	2	28	1	20	1	5	39	0	0
Cod liver	4	1	13	2	4	0	18	32	16	10

Suggest why the fish should have evolved with such a different pattern from that of the mammal. (Naturally occurring unsaturated acids are almost all *cis*-acids.)

There is an even more subtle consequence of the intake of unsaturated fats. Linoleic and arachidonic acids are the starting points for synthesis in the body of a series of compounds called prostaglandins, which have been linked with many body processes, including blood clotting (see Section 28: Headache or hangover? p. 105). One of the prostaglandins, thromboxane A_2, has a very powerful effect on blood platelets, causing them to clump together strongly. Another, prostacyclin, has exactly the opposite effect. The tendency of platelets to cluster together (and possibly to block an artery) depends on the balance between these two, and other, compounds. Perhaps the **type** and **amount** of fat that you eat alter the amounts of prostaglandins that are produced, and that in turn might affect blood clotting.

Faced with all this, what are the food manufacturers to do? The answer is clear: they maximise their profits. So butter is advertised as 'natural', despite the fact that it consists of 81% fat, most of it saturated. Margarine manufacturers stress how much polyunsaturated fat their product contains, despite the fact that the best advice is to *cut the intake of all fats*.

In fact, it is not clear how unsaturated a margarine can be. We saw above that polyunsaturated fatty acids tend to be liquid at room temperature, and to convert them to solids, some degree of saturation is required. This process is known as 'hardening' and usually involves reacting the oil with hydrogen at a pressure of 2–10 atmospheres and a temperature of 160–220°C in the presence of a nickel catalyst. By controlling the conditions, the manufacturer can saturate the required number of the carbon-carbon double bonds.

To check this, it is vital to determine the number of C=C groups in the molecule before and after hydrogenation. Nowadays, this is usually

CONTINUED

by chromatography, but originally it was done by reacting the fat with iodine. Any naturally occurring fat will be a mixture, and the extent of unsaturation will be expressed as the number of grams of iodine reacting with 100 g of fat. Thus butter might have an iodine value of 25–45 (that is to say, 100 g of butter will react with 25–45 g of iodine). A polyunsaturated oil like linseed oil will have an iodine value of 175–200.

Questions

4. Margarine is made by 'hardening' oils by reacting them with controlled amounts of hydrogen in the presence of a nickel catalyst. It appears that the hydrogen and the alkene are both strongly adsorbed on the nickel surface, *resulting in a weakening of the* $H-H$ *bond and of the* $C=C$ *double bond* (fig 4).

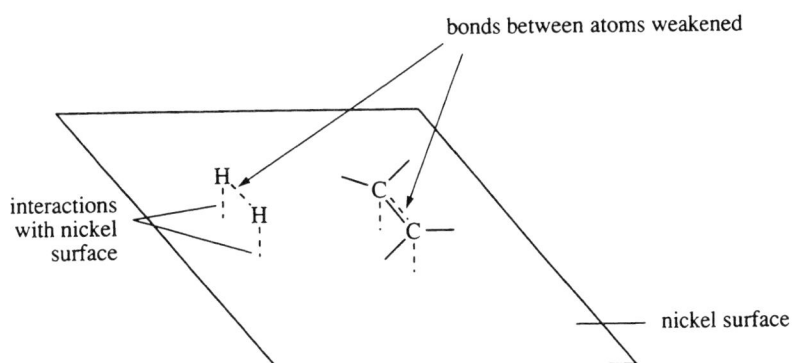

FIG 4

The hydrogens now attach themselves to the alkene, and the resulting alkane leaves the catalyst surface.

(a) Explain why this makes the oils more solid.

(b) When a *cis* alkene is exposed to a nickel catalyst, some isomerisation takes place. The *cis* isomer is converted into the *trans* form. Explain how this happens.

(c) What effect would you expect this isomerisation to have on the melting point of the fat?

(d) Saturated fats are more undesirable in the diet than *cis* mono-unsaturated ones. What about *trans* mono-unsaturated fats?

5. (a) Draw the structure of the product you would expect to be formed when oleic acid (9-octadecenoic acid) reacts with iodine.

(b) Draw the mechanism for the reaction of bromine with an alkene. Which is the most difficult step of the reaction?

(c) This mechanism is an *electrophilic addition*. Explain the meaning of the two words in italics.

(d) Bromine functions as an electrophile in this reaction. How does the bromine atom become an electrophile in the course of the reaction?

(e) The reaction of iodine with an unsaturated fat forms the basis of Wijs's method of estimating the amount of unsaturation in the fat. To get the reaction to occur rapidly, Wijs modified the reagent so that the attacking species is ICl (iodine monochloride). ICl is a better electrophile than bromine or iodine. Why is this? What effect will that have on the rate of the reaction?

(f) Calculate the iodine value of oleic acid ($C_{17}H_{33}COOH$). (Use your data book to find the relative atomic masses.)

21. Tonight, Josephine!

Unlike humans, other forms of life are not sexually receptive at all times. It is important, therefore, that the female of the species can signal to the male when she is ready to mate. Many animals do this visually – for example, baboons – but it is becoming increasingly obvious that chemical signals play a major part. These messenger molecules are called pheromones.

The role of pheromones is most clearly established for insects, where many different compounds are known. As well as sex attractants, there are pheromones with other functions. For example, some compounds cause insects to cluster together. This may be so that the discoverer of a good food source can attract others to it, or it may act to summon assistance if the nest or the queen is attacked. The pheromone is usually released from a gland at the tip of the insect's abdomen, and allowed to evaporate onto the breeze. The amounts released are small (1–2 *micro*grams is typical), but cases have been recorded in which males have been 'summoned' from a distance of up to five kilometres. Obviously the detectors must be very sensitive. Just one molecule is all that is needed to create an impulse in the receptor nerve of the silk worm moth.

So what?

With a world population approaching five thousand million, it is vital to limit the amount of our food taken by insects. Insects carry diseases, too; think of the malarial mosquito and the tsetse fly. Conventional insecticides have saved millions of lives, but unfortunately, they tend also to kill other organisms, many of them beneficial. Pheromones offer a much better alternative. Amounts are much smaller, for one thing, and they will presumably act only on the target species.

Their use varies. A small amount of attractant can be employed to see if there are any of the pests in the area, or to detect when they arrive, so that conventional insecticides can then be used. Pheromones are used to attract the males to a trap, or to interrupt mating behaviour. Unfortunately, pheromones are quite big molecules, and may be most effective in a complex mixture. Synthesis of these molecules is not easy, and to get them pure and cheap enough for farmers to use is a major challenge for the chemist.

Communication by pheromones doesn't end with insects. Look at a dog near a lamppost, for example. And the role of pheromones in the mating of pigs has been known for some time. When she is ready to mate, the sow adopts a special stance so that the boar can mount her. A mixture of compounds (such as [2] and [3] in fig 1) causes her to do this. The molecules are quite similar in shape to the human male sex hormone testosterone, and are present in the boar's saliva, sweat and urine.

They are so effective that they are now available in an aerosol. When a sow is in season and the breeder wants her to mate, a quick squirt puts her into the mating stance, ready for the boar.

More controversial is the question of human pheromones, but there is good evidence that they exist. To start with, we have glands in our skin which produce some of the same pheromones as pigs. These glands are concentrated in our armpits and pubic region. Perhaps the hair serves to disperse the scents like the wick on an oil lamp? Nor are we indifferent to each other's smell, although our response to it varies with time and culture. Most people nowadays find a clean body much

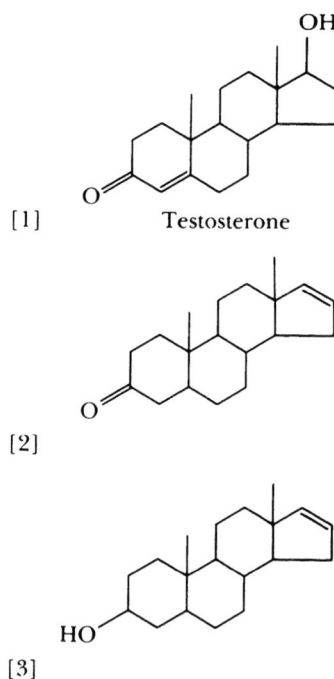

[1] Testosterone

[2]

[3]

FIG 1

CONTINUED

Sand castles and mud huts © 1991 Jeffrey Hancock, published by Hodder and Stoughton Educational

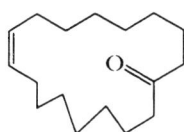

[4]   Civetone

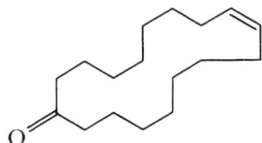

[5]

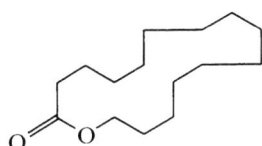

[6]

FIG 2

more alluring than a mucky one, but it wasn't always like that. After the battle of Austerlitz, Napoleon is supposed to have written home to Josephine, 'Ne te lave pas! Je reviens.' ('Don't wash! I'm coming home.')

There is more specific evidence for human pheromones. In one experiment, 60% of mothers could identify their new-born baby simply by smelling its head. (The fathers couldn't.) Very early in life babies learn to find their mother's breast, but not anyone else's, again, apparently, purely by smell. The menstrual cycles of women living together – sharing a room at college, for example – tend to become synchronised. This has been shown to be caused by a chemical secreted by one of them.

Experiments looking for a human sex attractant have not been conclusive, but the compound civetone (a cyclic compound of formula $C_{17}H_{30}O$, from a gland of the civet cat) has been used for centuries in perfumes. The molecule is flexible and can adopt many different shapes (see [4] and [5] in fig 2). Is it an accident that one of these [5] has quite a similar shape to testosterone?

There is another compound – [6] in fig 2 – with a musk-like perfume, which has a much stronger scent for women than for men. Interestingly, women are most sensitive to it when they ovulate, at precisely the time in their monthly cycle when they are most fertile.

The most intriguing fact of all, perhaps, is seen when we test the smell of the boar sex attractants mentioned above on humans. Not everybody can smell anything, and of those people who can, not everyone likes the odour, but a sizeable proportion of the humans tested can smell something, and find the smell appealing.

Why do we use artificial perfumes? Because they smell like sex attractant molecules?

Questions

On halogenoalkanes, useful synthetic intermediates

1. Pheromones are almost invariably quite large molecules.

    (a) Why would small molecules be less satisfactory? (Remember that a pheromone must act on only one species.)

    (b) Why would very large molecules not be satisfactory?

2. A chemist prepared molecule A (fig 3 on the next page), and intended to convert it to compounds B and C. (C is one of the boar sex hormones, and B is very similar to testosterone.) The **bold** letters are to help you focus on the important bits of the molecules.

    (a) A, B and C are ketones; they have a C=O group. What *other* functional group is present in (i) B and (ii) C?

    (b) A is also a chloroalkane. Chloroalkanes are said to be polar. What does this mean? Why are they polar?

    (c) This polarity means that halogenoalkanes are easily attacked by a certain *type* of reagent. What type? Why?

    (d) What *sort* of reaction is involved in the conversion of A to B?

    (e) Suggest suitable reagents and conditions for carrying it out.

    (f) Propose a mechanism for the reaction.

    (g) What *sort* of reaction is involved in the conversion of A to C?

    (h) Suggest suitable reagents and conditions for carrying it out.

    (i) Propose a mechanism for the reaction.

CONTINUED

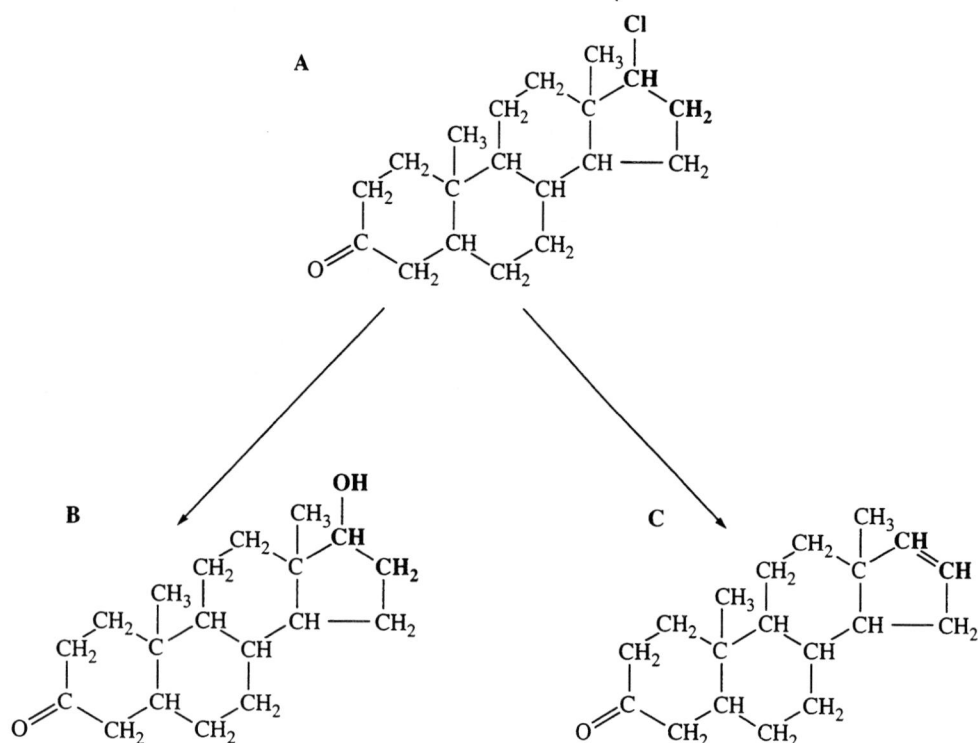

FIG 3

3. The major component of an attractant for corn earworm is
cis-11-hexadecenal. The final stages of a synthesis of this are given in
fig 4. Again, the **bold** type is to help focus your attention on the
important bits.

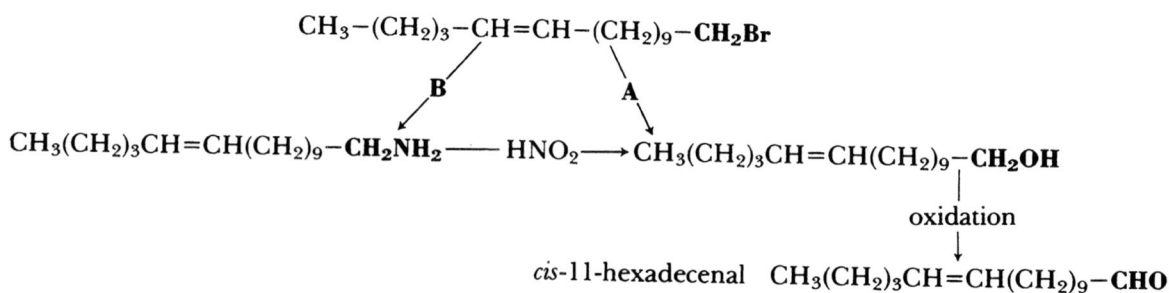

FIG 4

The chemist is uncertain whether to use path **A** or path **B**. Although **B**
involves an extra step, she wonders if it might give a better yield than **A**.

(a) What reagent and conditions would you use to carry out **A**?

(b) What reagent and conditions would you use for **B**?

(c) What problems are there with this reaction? How could you
minimise them?

(d) Advise the chemist. Which pathway, **A** or **B**, would you expect to
give the better yield?

(e) Someone suggests that it would be better to start with the iodide,
$CH_3(CH_2)_3CH=CH(CH_2)_9CH_2I$, rather than the bromide. What do
you think? Would it react more rapidly or give a better yield? Why?

CONTINUED

Sand castles and mud huts © 1991 Jeffrey Hancock, published by Hodder and Stoughton Educational

4. Muscalure (*cis*-9-tricosene) is a housefly sex attractant. A possible synthesis of this molecule ends with the step:

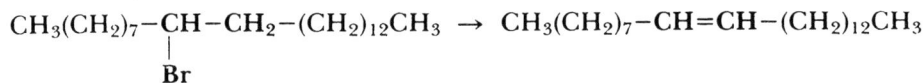

$$CH_3(CH_2)_7-\underset{\underset{Br}{|}}{CH}-CH_2-(CH_2)_{12}CH_3 \quad \rightarrow \quad CH_3(CH_2)_7-CH=CH-(CH_2)_{12}CH_3$$

(a) What does the '9' mean in the name of the compound?

(b) What *sort* of reaction is this?

(c) Suggest a suitable reagent and conditions that you might use to carry it out.

(d) There are at least four other products that could result from your reaction. How many can you think of?

(e)* What is meant by *cis* in the name of the compound?

(f)* Suggest why it is so crucial that the pheromone be the *cis* form.

5. Valeric acid ($CH_3(CH_2)_3COOH$) is a sex attractant for a wire worm that attacks sugar beet. The following synthesis from butan-1-ol has been proposed:

$$CH_3CH_2CH_2CH_2OH \rightarrow CH_3CH_2CH_2CH_2Cl \rightarrow CH_3CH_2CH_2CH_2CN \rightarrow CH_3CH_2CH_2CH_2COOH$$

(a) Suggest how you could carry out the first two steps of this reaction. Mention any special conditions.

(b) The $-CN$ group in the nitrile, $CH_3CH_2CH_2CH_2CN$, is polar. Why is it polar? Show the direction of the polarity by a suitable diagram.

(c)* The last step of this sequence is carried out by boiling the nitrile with water (and a catalyst). Can you use your answer to (b) to suggest how the water attacks the $-CN$ group?

(d)* Valeric acid is the old name for the product. Can you name it more systematically?

22. Ales and lagers

Beers are made from four basic ingredients; malt, hops, water and yeast. Malt is the name given to barley seeds which have been allowed to start germinating, until the starches of the seeds have been broken down by enzymes into the sugars needed for fermentation. The germination is then stopped, usually by drying.

The next stage is mashing. The malt is ground up and mixed with hot water, and a small amount of unmalted barley and sugar may be added. The mixture is kept at 65°C for up to several hours. Many different substances, including a wide variety of sugars, dissolve into the water, producing a liquid called 'wort'.

The wort is now boiled with the hops for 90 minutes or so in a large copper vessel. This extracts the bitter-tasting substances. After cooling, yeast is added and the wort fermented, for two to ten days.

The beer is now filtered and put into casks. A little more sugar is added, and it ferments further – there is always a little yeast left – and the resulting CO_2 dissolves in the beer under pressure, producing the fizz.

A simple process, and one practised for several thousand years, but it produces a varied and complex mixture. Over 400 different compounds have been identified in an average pint.

The major component (after water) is ethanol (usually just called 'alcohol'). The bitter flavour of the beer comes from compounds called humulones extracted from the hops. The other major flavour component is the dissolved carbon dioxide. In addition to these, there are many compounds that contribute to the taste. These may be compounds that were present in the original barley or hops and remained unchanged throughout the whole process. Alternatively, they may have been formed by reactions during the production of the beer. For example, there are other alcohols, such as propan-1-ol and propan-2-ol, three of the possible isomeric C_4 alcohols, three more C_5 alcohols, and so on. Many of these are present in concentrations too low to taste (although 2- and 3-methylbutan-1-ol are just above the taste threshold). There are the oxidation products of the alcohols also: aldehydes, ketones and carboxylic acids. The average ethanal concentration is below eight parts per million, but if it rises above about ten parts per million, the beer starts to taste stale. Ethanoic acid has a similar effect. It occurs in lagers at 50–150 parts per million. If it rises above 175 parts per million, the lager tastes vinegary. Then there are esters, several of which occur in concentrations high enough to taste.

It is the alcohol in the beer which will affect us most when we drink it. What happens?

Ethanol is readily absorbed, about 20% of it through the stomach, the rest from the intestine. It reaches the bloodstream – and therefore the brain – within five minutes of being drunk, but its maximum effect may not be felt for up to two hours. It is broken down in the body by enzymes in the liver, first to ethanal, then to ethanoic acid, and ultimately to CO_2 and water. Methylated spirit is ethanol which has had various substances added to it to make it unfit to drink. The chief one of these is methanol. This is especially damaging, as it is oxidised to methanal which reacts with many body chemicals, particularly proteins. The proteins of the optic nerve seem to be the most sensitive to methanal, and drinking meths regularly causes blindness.

The effects of ethanol on the body are very complex. One of the most

CONTINUED

Sand castles and mud huts © *1991 Jeffrey Hancock, published by Hodder and Stoughton Educational*

obvious is that it makes you urinate. This is not only because of the quantity of fluid you have drunk, but also because ethanol reduces the production of a hormone called ADH. The function of ADH is to control the extent of urination, so if its concentration falls, you urinate more. (This may be one of the causes of hangover: you lose so much water that you become dehydrated. Drinking several glasses of water may make you feel better next day.)

How ethanol affects the brain is still imperfectly understood, but its mechanism is probably like this. The brain and indeed the whole nervous system consists of millions of nerve cells, and when we think or feel or do anything, nerve impulses have to be passed from one nerve cell to the next. The message passes along one nerve cell by an electrical discharge, but it gets to the next cell by chemical means. A compound is released from the end of one nerve cell, diffuses across the gap to the next cell, and triggers the electrical discharge in this second cell, and so on.

The ethanol binds to one end of the nerve cells, making the chemical messenger molecules unable to trigger the electrical discharge, and so blocking the nerve impulse. Thus ethanol acts to depress brain activity. It seems that it first affects those centres of the brain responsible for behaviour control and so releases your inhibitions. It also affects your ability to carry out physical operations. One pint of beer before driving approximately doubles your chance of a crash (although you are probably still well below the legal limit).

For these reasons there are legal limits to the amount of ethanol you may have in your blood and still be allowed to drive (80 mg per 100 cm^3 of blood in Great Britain), and that in turn has led to a demand for beers which are low in ethanol.

There are essentially two ways of producing a low alcohol beer. One is to stop the fermentation before the ethanol concentration is too high (usually by cooling the wort to 0°C). The other is to ferment the wort normally and to remove the ethanol later, for example, by distillation.

Questions

On the chemistry of alcohols

1. There are four possible isomeric saturated C_4 alcohols, three of which occur in beer.

   (a) What is meant by 'isomeric'?

   (b) Draw their structures and name them.

2. Pentan-1-ol and hexan-1-ol are both present in beer (although in small amounts: the concentration of pentan-1-ol, for example, is about 0.2 ppm).

   (a) Why are they soluble in beer, which is mostly water?

   (b) They are much less soluble in water than methanol or ethanol are. Why is this?

   (c) What happens to their solubility in water if they are converted into esters? Why?

3. (a) What reagents would you use to convert ethanol to ethanal in the laboratory?

   (b) Devise an ionic equation.

   (c) The major problem with this reaction is the further reaction of the ethanal to ethanoic acid. How would you avoid this or, at least, minimise it?

CONTINUED

(d) In the early days of the drink-driving laws, the amount of ethanol in a driver's breath (and blood) was determined by a device called the 'breathalyser'. The driver blew into a bag through a tube containing orange crystals of potassium dichromate(VI), moistened with acid. The amount of green colour produced in the tube was a measure of the amount of ethanol in the breath. Explain how it worked.

4. (a) Suggest why the concentrations of ethanal and ethanoic acid increase when a beer is left to stand in air.

(b) Devise an equation for the conversion of ethanol to ethanal.

(c) What would happen under these conditions to decan-2-ol, with a concentration in beer of 0.005 ppm?

(d) The structure of one of the humulones in beer is given in fig 1.

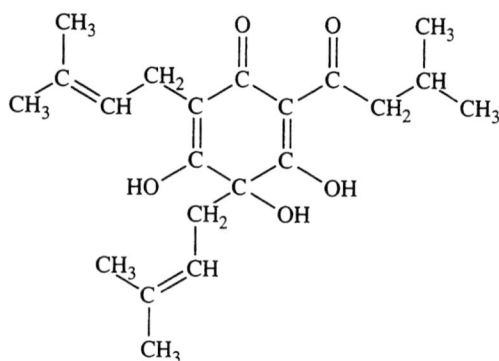

FIG 1

Notice that it is also an alcohol: there are three —OH groups in the molecule. Why won't the humulone react like ethanol, when beer is left to stand in air?

(e) The concentrations in beer of some compounds containing *nine* carbon atoms are (all in ppm): alcohols 0.11; aldehydes 0.0038; ketones 0.030; carboxylic acids 6.88; esters 0.18. Suggest why the concentration of aldehydes is so low.

5. Ester concentrations are quite high. Thus ethyl ethanoate can occur up to levels of 70 parts per million, especially in stout. (Its taste threshold is 33 ppm.)

(a) Write an equation for its formation.

(b) Its formation during fermentation takes 30 hours or so. The laboratory preparation of the ester takes about two hours. How is it speeded up so much?

6. The pH of beer is typically about 3.9–4.0. Why do you think it is so low?

7. The traditional method of specifying the strength of a beer was the 'original gravity' (OG). This is the density of the original wort before fermentation, multiplied by 1000. (It is sometimes given on the bottle or can.)

(a) What would be the OG of a beer made from a wort of density $1.042 \, \mathrm{g \, cm^{-3}}$?

(b) Why is the wort more dense than water?

(c) What will happen to the density of the wort as it is fermented? Why?

CONTINUED

(d) The original gravity and ethanol concentration of several beers are given in table 1.

TABLE 1

Beer	Original gravity /OG	Ethanol /%
John Smith's Strong Ale	1045–1051	5.0
John Smith's Bitter	1034–1038	3.8
John Smith's Magnet	1038–1042	4.0
John Smith's Magnet Old	1068–1074	6.5
Courage Russian Stout	1098–1104	10.0
Courage Light Ale	1030–1034	3.2
Courage Bulldog Pale Ale	1065–1071	6.3
Theakston Old Peculier	1055–1059	5.6
Theakston XB	1042–1046	4.5
Theakston Best Bitter	1036–1040	4.2

Is the OG related to the ethanol content? How? Why? (Plot a suitable graph.)

(e) According to *The Guinness Book of Records*, the strongest beer in the world is reputed to be one called 'Roger and Out', brewed by the Frog and Parrot pub in Sheffield, with an OG of 1125. What do you think is its percentage of ethanol?

(f) 'Miller Lite' beer is brewed to be low in carbohydrate (and therefore less fattening, although ethanol itself is fattening). Its OG is 1030–1034, and its alcohol content 4.2%. Is it low in carbohydrate?

(g) In Britain (and Belgium) the duty payable on a beer used to be determined by its OG value. In other European countries, it is determined by alcohol content. Which do you think is better and why?

23. Toffee and toast

TOFFEE

250 g butter
1 kg soft brown sugar
100 cm³ vinegar
200 cm³ water

*Put the butter, soft brown sugar, vinegar and water into a large
saucepan. Boil slowly for fifteen minutes. Test by dropping a little
of the mixture into cold water. When it goes crisp, remove from
heat. Pour into a shallow oiled tin, leave to set, then break it into
pieces when it is cold.*

The reaction of sugars on heating to produce this thick brown
concoction depends on some complex but very common chemistry.
(Think how many cooked foods are brown in colour: toast, meat, cakes,
bread, chips . . . even beer is brown. It's all the same sort of reaction.)

The first step is an isomerisation. For example, glucose, $C_6H_{12}O_6$,
reacts as shown in fig 1. (Don't be put off! It's not actually as bad as it
looks at first. The bold type is to help you to focus on the important bits
of the molecules.)

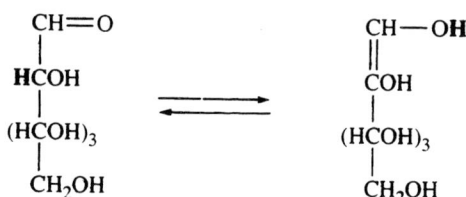

CH=O   CH—OH
 | ||
HCOH   ⇌   COH
 | |
(HCOH)₃   (HCOH)₃
 | |
CH₂OH   CH₂OH

FIG 1   glucose

This reaction is then followed by several more steps, with the loss of
two molecules of water and the formation of a 3,4-dideoxyosone. This
compound then reacts as shown in fig 2.

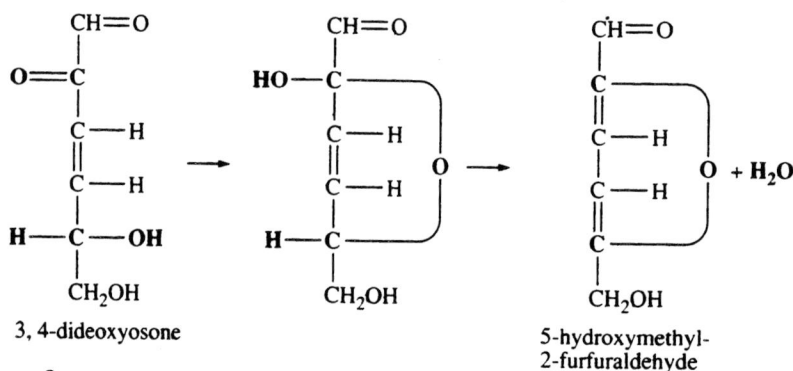

CH=O   CH=O   CH=O
 | | |
O=C   →   HO—C   →   C
 | | ||
C—H   C—H   C—H
‖   ‖ |
C—H   C—H O   C—H   O + H₂O
 | | ‖
H—C—OH   H—C   C
 | | |
CH₂OH   CH₂OH   CH₂OH

3, 4-dideoxyosone   5-hydroxymethyl-
                                  2-furfuraldehyde

FIG 2

The 5-hydroxymethyl-2-furfuraldehyde undergoes polymerisation,
resulting in large molecules with the characteristic brown colour.

CONTINUED

Sand castles and mud huts © 1991 Jeffrey Hancock, published by Hodder and Stoughton Educational

Questions

1. What is meant by:

   (a) 'isomerisation'?

   (b) 'polymerisation'?

2. Most of these reactions involve a nucleophilic attack on the carbonyl group.

   (a) What is a nucleophile?

   (b) Why are carbonyl groups susceptible to nucleophilic attack?

3. Ethanal undergoes a nucleophilic addition with HCN.

   (a) What is the nucleophile here?

   (b) Give an equation for the reaction.

   (c) Draw the mechanism of the reaction.

4. The first step in fig 2 is also a nucleophilic addition.

   (a) What is the nucleophile here?

   (b) Draw the mechanism of the reaction. (Use your answer to question 3 to help you.)

In the presence of amino compounds, the browning of sugars occurs much more rapidly, by the Maillard reaction (named after the Frenchman who discovered it in 1912).

The aldehyde group of the sugar reacts with the amino group to give a Schiff base (fig 3).

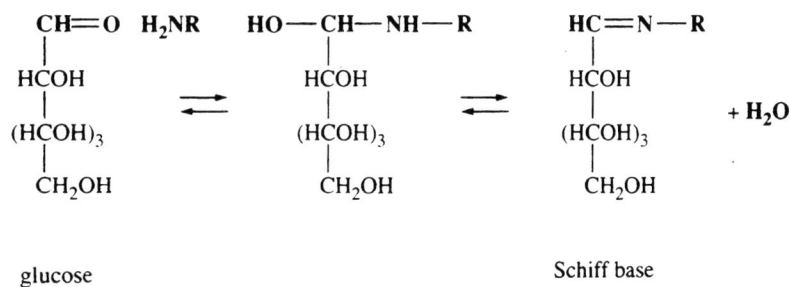

$$
\begin{array}{ccc}
\text{CH}{=}\text{O} \;\; \text{H}_2\text{NR} & \text{HO}{-}\text{CH}{-}\text{NH}{-}\text{R} & \text{HC}{=}\text{N}{-}\text{R} \\
| & | & | \\
\text{HCOH} & \text{HCOH} & \text{HCOH} \\
| & | & | \\
(\text{HCOH})_3 & (\text{HCOH})_3 & (\text{HCOH})_3 \quad + \text{H}_2\text{O}\\
| & | & | \\
\text{CH}_2\text{OH} & \text{CH}_2\text{OH} & \text{CH}_2\text{OH}
\end{array}
$$

glucose   Schiff base

FIG 3

The Schiff base now undergoes a reaction known as the Amadori rearrangement (fig 4).

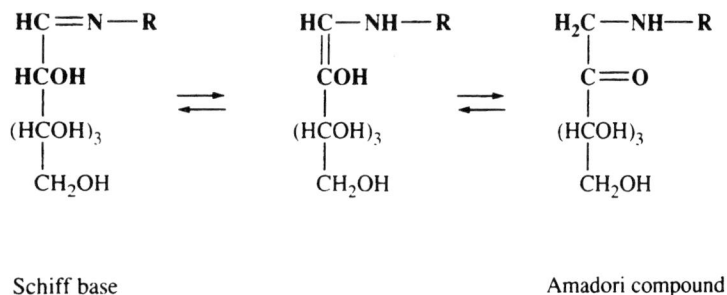

$$
\begin{array}{ccc}
\text{HC}{=}\text{N}{-}\text{R} & \text{HC}{-}\text{NH}{-}\text{R} & \text{H}_2\text{C}{-}\text{NH}{-}\text{R} \\
| & \| & | \\
\text{HCOH} & \text{COH} & \text{C}{=}\text{O} \\
| & | & | \\
(\text{HCOH})_3 & (\text{HCOH})_3 & (\text{HCOH})_3 \\
| & | & | \\
\text{CH}_2\text{OH} & \text{CH}_2\text{OH} & \text{CH}_2\text{OH}
\end{array}
$$

Schiff base   Amadori compound

FIG 4

CONTINUED

The central nitrogen-containing compound of fig 4 can react further; in particular, it can decompose, forming another, different, aldehyde (fig 5).

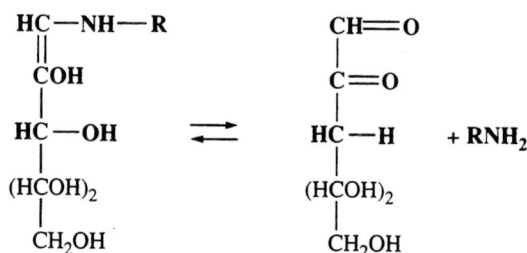

$$
\begin{array}{ccc}
\text{HC}-\text{NH}-\text{R} & & \text{CH}=\text{O} \\
\parallel & & \mid \\
\text{COH} & & \text{C}=\text{O} \\
\mid & & \mid \\
\text{HC}-\text{OH} & \rightleftharpoons & \text{HC}-\text{H} \qquad + \text{RNH}_2 \\
\mid & & \mid \\
(\text{HCOH})_2 & & (\text{HCOH})_2 \\
\mid & & \mid \\
\text{CH}_2\text{OH} & & \text{CH}_2\text{OH}
\end{array}
$$

FIG 5

This compound reacts to form a 3,4-dideoxyosone. This in turn is converted to 5-hydroxymethyl-2-furfuraldehyde (fig 2). The formation of the brown-coloured polymers then follows.

Questions

5. Aldehydes and ketones react with 2,4-dinitrophenylhydrazine, forming an orange precipitate.

   (a) Write the equation for the reaction between ethanal and 2,4-dinitrophenylhydrazine.

   (b) The first step of this reaction is also a nucleophilic attack. What is the nucleophile?

   (c) Draw the mechanism of the reaction.

   (d) Draw the structure of the product formed when 2,4-dinitrophenylhydrazine reacts with glucose.

6. The first reaction in fig 3 is also a nucleophilic addition.

   (a) Explain what the nucleophile is.

   (b) Draw the mechanism of the first reaction in fig 3.

   (c) This reaction will not work in an acid solution because the amino group becomes protonated:

$$\text{RNH}_2 + \text{H}^+ \rightarrow \text{RNH}_3^+$$

Why does this prevent the reaction from occurring?

   (d) In the toffee recipe, one ingredient is vinegar (ethanoic acid solution). Why won't the addition of this acid prevent the reactions in fig 3 occurring?

7. The reactions in figs 1 and 2, and those in figs 3–5 both produce the compound, 5-hydroxymethyl-2-furfuraldehyde. If they both result in the same end product, what is the function of the amino compound, RNH_2?

These reactions are responsible for much of the flavour and colour of our food, but there are two occasions when they can cause problems.

When a foodstuff containing protein and carbohydrate is stored for a long period, the amino groups in the proteins may react with sugars (figs 3–5). The reaction may go no further than the Amadori compound (fig 4), so there may be no obvious browning, but much of

CONTINUED

Sand castles and mud huts © 1991 Jeffrey Hancock, published by Hodder and Stoughton Educational

the amino acid may be bonded to sugars in this way. This makes the amino acid unavailable for digestion, so decreasing the nutritional value of the food. This can be a serious problem for countries using food stores to supply emergency relief in disaster areas.

The other problem with the Maillard reaction occurs in the body itself. In the presence of high concentrations of sugar – which commonly occur in people with diabetes, or if someone eats a sugary snack on an empty stomach – protein molecules react with the sugar. This reaction – glycation – alters the properties of the protein. The effect of this high sugar concentration may last long after the snack has been digested. The Schiff base (fig 3) is produced in a matter of hours after the sugar is eaten, but the Amadori product (fig 4) may take weeks to form and many weeks more to react with protein. These effects of the sugar are undone only if the protein or Amadori compound are broken down by the body, and some proteins are very long lived indeed. Among these long-lived proteins are the crystallins of the eye. Glycated molecules of crystallin clump together to give opaque suspensions. This is perhaps why cataracts are five times more common in people with diabetes.

It is possible that protein glycation is a cause of the ageing of the body. And since high concentrations of sugar in the blood cause glycation of proteins, eating a bar of chocolate on an empty stomach may just be a quick way to get old. . . .

Questions

8. Boiled food, or food cooked in a microwave oven, tends not to be very brown. Roasted or fried food, cooked at a higher temperature, is much browner. Why is this?

9. Look up the structure of the amino acid lysine and draw the structure of the Amadori compound it will form with glucose.

10. Suggest how the Amadori product in fig 4 could link proteins together by reacting with another amino acid.

24. Wrinkles and waves

You have probably seen the preparation of nylon in the laboratory. A solution of adipyl chloride (hexanedioyl chloride) in a suitable solvent (such as 1,1,1-trichloroethane) is poured into a beaker. On top of it is poured a solution of 1,6-diaminohexane in water. The two do not mix, and the aqueous solution floats on the top. Where the two layers meet, the reaction occurs, and a continuous thread of nylon can be pulled out. Industry uses the diacid rather than the dichloride for the reaction. The key idea is the same: monomers with two functional groups will link up to form extended polymers.

$$HOOC-(CH_2)_a-COOH + NH_2-(CH_2)_b-NH_2 \rightarrow -[CO-(CH_2)_a-CO-NH-(CH_2)_b-NH]_n- \quad [1]$$

The size of the monomers, indicated by the values of a and b, can vary, giving rise to nylons with different properties.

Nylon is a synthetic version of the natural polymers called proteins. The proteins of the body are built from 20 amino acids, of general formula $NH_2-CH(R)-COOH$, which are joined head to tail to form a polyamide (fig 1).

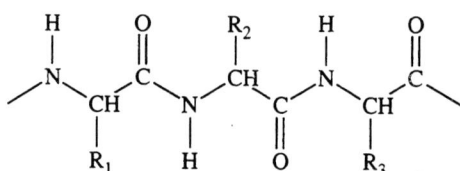

FIG 1

The side chains (indicated by R_1, R_2 and so on) can vary, from H (part of the amino acid glycine) to $-CH_2C_6H_5$ (phenylalanine), from $-CH_2COOH$ (aspartic acid) to $-(CH_2)_4-NH_2$ (lysine). The varying side chains confer properties which vary from protein to protein and which differ from those of nylon, of course. Let us focus on just one protein, α-keratin, found in hair and wool. This exists as a helical structure, with the $-(NH-CH-CO)-$ repeating unit forming a spiral with the R groups attached to it. Three of these α-helices wind round each other to form a protofibril. Eleven protofibrils make a microfibril and hundreds of these a macrofibril. Many macrofibrils make up a cell, and an individual hair contains many cells.

Questions

On carboxylic acids and their derivatives

1. When nylon is formed in the laboratory, the essential reaction is:

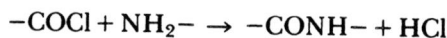

$$-COCl + NH_2- \rightarrow -CONH- + HCl$$

   (a) The NH_2- group is a nucleophile. What does this mean?

   (b) Why is the $-COCl$ group susceptible to nucleophilic attack?

   (c) Draw the mechanism of the reaction.

CONTINUED

Sand castles and mud huts © 1991 Jeffrey Hancock, published by Hodder and Stoughton Educational

(d)* Aldehydes are also susceptible to nucleophiles containing the NH_2- group, for example, hydrazine, NH_2NH_2. Draw the mechanism of this reaction.

(e)* Explain why these two mechanisms take such a different course after the same first step.

2. Industrially, the diacid is used; in the laboratory, the dioyl chloride is used because the chloride reacts faster.

(a) Why isn't the dioyl chloride used by industry?

(b) How could you convert the diacid into the dioyl chloride in the laboratory?

3. Nylons produced by the process summarised in the equation [1] are named as 'nylon-(a + 2),b'. Thus the compound with the repeat unit $-[CO(CH_2)_4CONH(CH_2)_6NH]-$ is called nylon-6,6.

(a) What do these numbers mean?

(b) Name the nylon with repeat unit:

$$-[CO(CH_2)_2CONH(CH_2)_6NH]-.$$

(c) Draw the repeat unit of nylon-6,10.

Nylons and proteins are both amides, and so have some similar properties. Both can be hydrolysed back to the original acid and amine, the reaction being catalysed by any acid or base. That is why nylon and silk – made of the protein, β-keratin – are slowly decomposed by the 2-hydroxypropanoic acid in sweat.

The $-CO-$ and $-NH-$ groups are polar, so in nylon and proteins hydrogen bonding occurs between these groups, either between different parts of one molecule or holding different polymer molecules together. This is how the nylon molecules are linked, contributing to its strength. Hydrogen bonds also help to determine the shape of protein molecules. In hair, for example, they hold the turns of the helical structure together. This also explains why both nylon and proteins attract water: the water molecules hydrogen bond to the $-CO-$ and $-NH-$ groups, but since only these small parts of the molecule interact with the water, nylon absorbs little of it. Contrast this with cotton, which is a polysaccharide with many $-OH$ groups in the molecules, all of which can attract water. In hot weather, cotton shirts can absorb a lot of sweat without becoming wet, so they are comfortable to wear. Nylon, on the other hand, can't absorb much sweat, so nylon clothing feels sticky and uncomfortable. Proteins can absorb still more water, which is why woollen socks are best if you are going to walk a long way and why your skin wrinkles when you laze too long in the bath.

And haven't the Caucasians amongst you gone to bed at some time with wet hair and found next morning silly little curls sticking out from your head? Hydrogen bonding again – as the water evaporated from your hair, the protein chains formed hydrogen bonds with each other, fixing them in the new 'style'. Fortunately, if you wet your hair again, the unwanted hydrogen bonds are broken. You can now restyle your hair before allowing hydrogen bonds to reform again as it dries correctly. (Those of you who are black are free of this embarrassment: your hair isn't altered by water.)

Changes in the properties of nylons can be made by alterations in the nylon chain. A major use of nylon now is in bearings and gears, where the nylons are made from longer chain monomers, so that water absorption is reduced still further. If the flexible compounds of equation [1] are replaced by much more rigid monomers (fig 2), the resulting nylon is very hard (nine times harder than steel) and is used in car tyres and bulletproof vests.

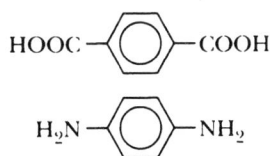

$HOOC-\langle\bigcirc\rangle-COOH$

$H_2N-\langle\bigcirc\rangle-NH_2$

FIG 2

CONTINUED

Unlike nylon's, the protein backbone does not change, but the side chains can vary, and these changes have their own effects. There is more in the next section on perms and hair conditioners. (See Section 25: Shampoo and set, p. 93.)

Questions

4. (a) What is meant by a 'hydrogen bond'?

(b) How do the C=O and N−H groups in proteins and nylons form hydrogen bonds?

(c) Redraw fig 1, then draw a second chain to show how the two chains hydrogen bond to each other. [*Hint*: Draw the two chains in opposite directions.]

5. Like all amides, proteins and nylons can be hydrolysed.

(a) What is meant by hydrolysis?

(b) Write an equation for the hydrolysis of nylon-6,6.

(c) Draw the first step of the mechanism, showing how the H_2O molecule attacks the nylon chain.

(d) If the reaction is done with an alkali, it is faster. Why do OH^- ions attack faster?

(e) The hydrolysis is also catalysed by H^+ ions. Suggest how the H^+ ions could attack the nylon chain so as to speed up the hydrolysis.

6. All amides can absorb water.

(a) Suggest how the water is bonded to the amide. Draw a diagram to illustrate this.

(b) When exposed to air of 65% humidity, human hair can absorb 13% of water by mass, while nylon-6,6 can absorb only 3–4%. Why?

(c) 'Always cut your toenails after a bath; they're softer then'. Why should they be softer? Can you think of an experiment you could do to support your idea?

(d) Clothes are best ironed by passing a hot piece of flat metal over the slightly damp material. The heat dries the clothes, which come out flat. How does this work?

7. Polythene can be thought of as having the structure $-(CH_2-CH_2)_n-$. It softens at 110–150°C, the temperature depending on how it has been made. Nylon-6,6 melts at about 270°C. Explain the difference.

8. Different nylons are synthesised with different lengths of hydrocarbon chain; i.e. different values for a and b in: $-[CO-(CH_2)_a-CO-NH-(CH_2)_b-NH]_n-$. What do you think will be the effect of increasing a and b values on:

(a) the ability of the nylon to absorb water

(b) the melting point of the nylon

(c) the strength of the nylon

(d) the resistance of the nylon to hydrocarbon solvents?

9. Afro-Caribbean hair is naturally more tightly curled than the hair of Caucasians. Water doesn't wet it in the same way. After swimming or showering, for example, a couple of shakes may be all that is needed to dry it. Can you suggest why? (The chemical composition of each of the two sorts of hair is not significantly different.)

25. Shampoo and set

Hair and wool are both types of protein called α-keratins. Chemically, they are polyamides, with a backbone of:

$$-CO-CH-NH-CO-CH-NH-$$
$$\quad\quad\;\; |\quad\quad\quad\quad\quad\; |$$
$$\quad\quad\; R_1 \quad\quad\quad\quad\; R_2$$

where R_1 and R_2 can be any of about 20 different groups. The nature of these groups determines the properties of the protein. In hair and wool the most common of these include $-CH_2CH_2COOH$ (from the amino acid glutamic acid), and $-CH_2CH_2CH_2NHC(NH_2)=NH$ (from arginine). These are polar side chains, capable of hydrogen bonding. This is important in determining the shape of the fibres.

Hair dirt probably contains two main components. The major one is scalp oils, secreted from a gland at each hair root. Mixed with this will be dirt from the air, plus anything smeared or sprayed onto the hair, such as gel and setting sprays etc. Cleaning the hair involves getting all this into solution in water (or at least into suspension), using a shampoo. The principal component of shampoo is a detergent. This is a molecule with a large non-polar segment, and an ionic end. It will usually be an anionic detergent, such as sodium dodecyl sulphate:

$$CH_3-CH_2-CH_2-CH_2-CH_2-CH_2-CH_2-CH_2-CH_2-CH_2-CH_2-CH_2-O-SO_3^-\;Na^+$$

which can be represented as:

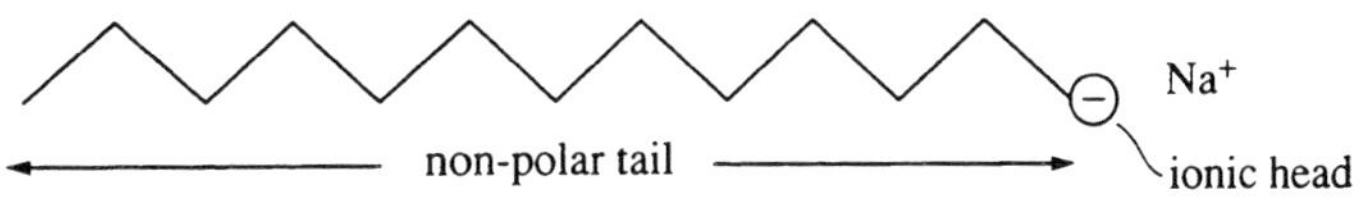

FIG 1

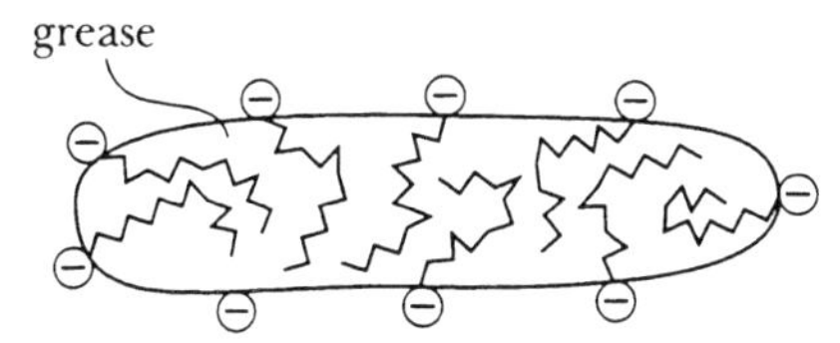

FIG 2

When this is rubbed onto the hair, the non-polar chains dissolve in the grease, while the ionic head remains in the water. Further massaging of the hair lifts the grease out into suspension in the water, surrounded by detergent molecules (fig 2). Repulsions between the negatively charged $-O-SO_3^-$ heads surrounding the grease prevent the droplets from joining up into one big greasy blob, and when you rinse the shampoo out, the grease is rinsed away with the water.

Questions

1. (a) Suggest what species will be formed in acidic solution from the side chains of:

(i) lysine, $-(CH_2)_4-NH_2$

(ii) arginine, $-(CH_2)_3-NH-C=NH$
$$\quad\quad\quad\quad\quad\quad\quad\quad\quad\quad |$$
$$\quad\quad\quad\quad\quad\quad\quad\quad\quad NH_2$$

(b) Suggest what will be formed in alkaline solution from the side chain of:

(i) aspartic acid, $-CH_2-COOH$
(ii) glutamic acid, $-CH_2-CH_2-COOH$.

CONTINUED

(c) The pH at which the number of positive charges on the side chains of the protein exactly equal the number of negative charges on the side chains is called the *isoelectric point*. The isoelectric point of hair is about 3.5–4.0. Under more alkaline conditions (such as at the pH of shampoos, about 6.5) the hair is negatively charged. Explain why.

2. (a) Shampoos contain anionic detergents like sodium dodecyl sulphate (fig 1). What happens to the Na^+ ions when this is added to water?

(b) The long hydrocarbon chain of sodium dodecyl sulphate dissolves in grease (fig 2). Why? What sort of interactions are occurring to encourage this to happen?

(c) The ionic head of the molecule, the $-O-SO_3^-$ group, interacts with water. How?

(d) Soaps have formulae such as $CH_3(CH_2)_{16}COO^-Na^+$. The hydrocarbon chain will dissolve in grease just like the detergent does. But the $-COO^-$ does not interact with water as well as the $-O-SO_3^-$ group does. Why not?

(e) In one shampoo test, a soap removed 29% of the grease from a hair sample, while a sodium dodecyl sulphate shampoo removed 51%. Explain the reason(s) for the difference.

3. (a) In the light of your answer to question 1(c), explain why shampoos containing anionic detergents are easy to rinse out of hair after shampooing at pH 6–7.

(b) After it is washed, Caucasian hair is impossible; it tends to fly up off your head. One reason is that the grease and dirt that was holding it down have been washed out. Can you suggest another?

The most common single amino acid in hair is cysteine, in which the R side chains consist of $-CH_2-S-S-CH_2-$, linking two different parts of the protein chain together, or even two different chains. It is these $-S-S-$ bridges that form the more permanent waves and curls, and it is these that are changed during a 'perm'. In the cold permanent waving process (invented in the early 1940s), a solution of a salt of thioglycolic acid ($HS-CH_2-COOH$), at a concentration of about 0.6 $mol\,dm^{-3}$ and at pH 9.0–9.5, is applied to the hair. This breaks around 20% of the $-S-S-$ bridges in the hair by a reaction of the type shown in fig 3.

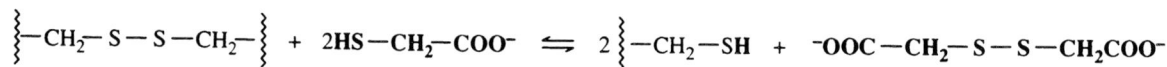

$$\}-CH_2-S-S-CH_2-\} + 2HS-CH_2-COO^- \rightleftharpoons 2\}-CH_2-SH + {}^-OOC-CH_2-S-S-CH_2COO^-$$

FIG 3

($\}$ stands for the keratin protein chain of the hair.) The hair is now restyled and left for 15–30 minutes (depending on hair type) to give the protein chains a chance to reorganise themselves into this new style. Finally, a weak oxidising agent, such as hydrogen peroxide, is used to reform the $-S-S-$ links, converting the $\}-CH_2-SH$ into $\}-CH_2-S-S-CH_2-\}$ and so holding the hair in the new style.

Brown hair is naturally coloured by melanins, red hair by trichosiderins. These are so complex that their exact structures are not known, but it is clear that the colour is due to the particular molecular structures. Bleaching destroys these structures by oxidising the molecules. The most commonly used oxidising agent is hydrogen peroxide, at pH 9–11.

CONTINUED

Sand castles and mud huts © 1991 Jeffrey Hancock, published by Hodder and Stoughton Educational

Hair may be artificially coloured (dyed). Permanent hair colouring is usually achieved by an oxidation dyeing process. A precursor, such as 1,4-diaminobenzene, is added to the hair and is absorbed onto its surface. A solution of hydrogen peroxide and a reagent called a coupler, such as 1,3-dihydroxybenzene can then be added. The hydrogen peroxide oxidises the precursor to a reactive intermediate, which combines with the coupler to form a light brown dye (fig 4).

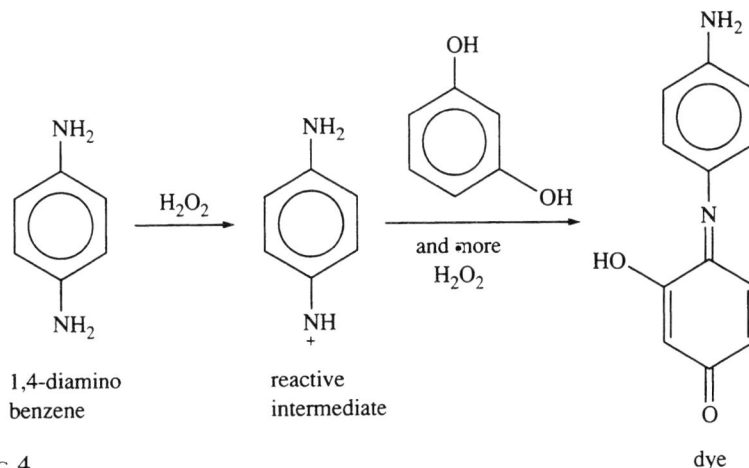

FIG 4

1,4-diamino benzene

reactive intermediate

dye

Further reactions give darker brown compounds. Other colours can be obtained by using different starting materials.

Finally, what about hair sprays, used to hold a style in place for sports or parties? Major components of these sprays are polymers which bind in various ways to the hair fibre and increase its 'body'. The forces binding the polymer to the hair can be identified by measuring the energy of the bonding. Some data is given in table 1. (Since the binding energy will be greater if the molecule is larger, the values quoted are for a length of the polymer of approximately 0.5 nm.)

TABLE 1

Polymer	Repeating unit	Binding energy /kJ mol^{-1}
polyethene	$-CH_2-CH_2-$	4
polybut-2-ene	$-CH_2-C(CH_3)_2-$	see question 7
polyethenol	$-CH_2-CH(OH)-$	20
polyamide	$-CH_2-CH-$ $\quad\quad\quad\vert$ $\quad\quad\quad CO$ $\quad\quad\quad\vert$ $\quad\quad\quad NH_2$	see question 7
polyetheneimine	$-CH_2-CH_2-\overset{+}{N}H_2-$	>400

Apparently, polyethene is bound to the hair by van der Waals' forces (dispersion or transient dipole-induced dipole forces). The other polymers are bound more strongly, polyetheneimine very much more so. This may be a problem, in that they may be very difficult to wash out of the hair.

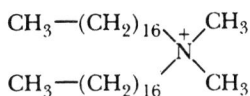

FIG 5

Questions

4. This question is about conditioners. You will need to use ideas and information from question 1. Conditioners contain small amounts of cationic detergents (such as fig 5) again in solutions of pH close to 7.

CONTINUED

(a) Explain why a cationic detergent is much more difficult to rinse out of your hair, if it has been washed at pH 6–7.

(b) Conditioners make the hair more manageable. It is easier to comb, rather than flying up all over the place. Why is this?

5. Suggest explanations for these observed effects of the permanent waving process on hair.

(a) It decreases the amount of cysteine and increases the amount of cysteine (R = $-CH_2-SH$).

(b) The amount of cysteic acid (R = $-CH_2-SO_3H$) is also increased.

(c) It makes the hair weaker.

6. Bleaches attack not only the coloured molecules but also the protein of the hair itself.

(a) Bleaches are used at pH 9–11. Suggest an effect that such a highly alkaline solution could have on the backbone of the protein (which is an amide, remember).

(b) The oxidising agent of the bleach (usually H_2O_2) could attack some of the side chains of the proteins. Suggest what effect an oxidising agent could have on:

(i) the $-CH_2-SH$ groups of cysteine

(ii) the $-CH_2OH$ of serine

(iii) the $-CH(OH)-CH_3$ of threonine.

7. This question is concerned with the binding of polymer molecules to hair.

(a) Polyethene molecules are bound to hair fibres using van der Waals' or dispersion forces. Using a simple example, explain how van der Waals' forces come about.

(b) Why is the binding energy of polyethenol so much higher than that for polyethene?

(c) Suggest values for the binding energies of polybut-2-ene and for the polyamide in the table. Explain your answers.

(d) Why does the polyetheneimine bind so strongly to the hair?

Sand castles and mud huts © 1991 Jeffrey Hancock, published by Hodder and Stoughton Educational

26. Flashes and bangs

Diesel oil burns. Mix it with a suitable oxidising agent and you turn it into an explosive. 2,4,6-trinitromethylbenzene is a runny liquid that burns if it is ignited. But if detonated, it explodes violently: it is TNT, still the most widely used military explosive.

What is the difference between a fuel – which burns smoothly, giving off heat – and an explosive?

It isn't the amount of heat given off. Explosives produce *less* heat than fuels. For example, TNT gives out $3428 \, kJ \, mol^{-1}$ when it burns, but when it explodes, only $953 \, kJ \, mol^{-1}$. It isn't the amount of gas produced either. When TNT explodes, there isn't time for oxygen in the air to react with it, so there isn't enough oxygen for complete oxidation. Less gas is actually produced than when it burns (see question 1).

No, the key thing that distinguishes an explosive from a fuel is its *speed* of reaction. If you put a fuel into a fuse and light it, the fuse would burn at one centimetre a second or less. If you could do the same with TNT, you would find that it burnt at *seven kilometres a second*! It is hardly surprising, then, that there is insufficient time for atmospheric oxygen to diffuse to the reaction mixture as it is required.

Of course, the explosive isn't actually burning. It may do so initially, perhaps for a microsecond or so, but the extremely rapid energy release produces a shock wave – a sudden increase in pressure – that causes the rest of the compound to decompose suddenly. Although the energy released is less than could be obtained from complete combustion, it is released so rapidly that very high temperatures and pressures are produced. The detonation pressure of TNT is about 2×10^{10} pascals – about 200 000 atmospheres.

Explosives are classified as primary or secondary, depending on their sensitivity to shock. Commercial explosives – military or industrial – will contain both. First there will be a small amount of the primary explosive, such as mercury fulminate ($Hg(ONC)_2$), lead azide ($Pb(N_3)_2$), lead styphnate or diazodinitrophenol (fig 1).

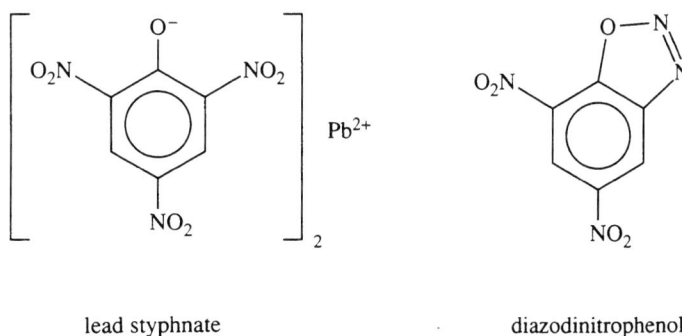

lead styphnate   diazodinitrophenol

FIG 1

This primary explosive is detonated easily, by ignition or some sort of mechanical shock such as a firing pin. The shock wave it produces (which may be 10^6 times greater than the initiating shock) sets off the secondary explosive. There is a wide range of these, including aliphatic nitro compounds like nitroglycerin or RDX (fig 2 on the next page).

CONTINUED

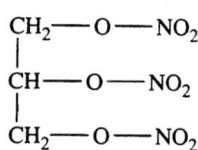

nitroglycerine   RDX

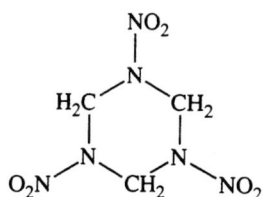

FIG 2

The majority are atomatic nitro compounds, for example picric acid or TNT (fig 3).

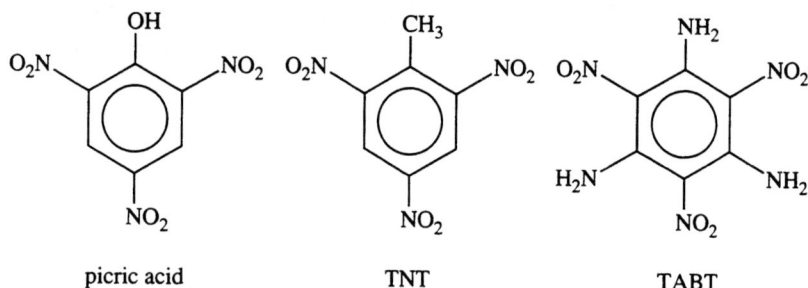

picric acid   TNT   TABT

FIG 3

What makes a good explosive? It must be an unstable compound, that is to say, its enthalpy of formation must be positive, so that energy is released when it decomposes. (The enthalpy of formation for mercury fulminate, for example, is $+64\,kJ\,mol^{-1}$). The problem with using unstable compounds, is that because they are unstable, they are liable to decompose spontaneously at any time. This is avoided by using compounds whose activation energy of decomposition is high. The higher the activation energy, the more stable the compound will be. So TNT has an activation energy of $143\,kJ\,mol^{-1}$, which makes it acceptably stable, but that for TABT is $250\,kJ\,mol^{-1}$. TABT is very safe to handle.

Questions

On reactions of aromatic compounds

1. (a) Devise an equation for the complete combustion of TNT.

(b) When TNT explodes, 1.21 moles CO_2, 1.98 moles CO, 1.60 moles H_2O, 1.42 moles N_2, 0.46 moles H_2, 0.16 moles NH_3 and 0.10 moles CH_4 are formed, per mole of TNT. What else is produced?

(c) What would you expect to *see* if a small amount of TNT was detonated?

2. TNT is manufactured by a variety of processes, but all of them involve the nitration of methylbenzene.

(a) What reagent(s) is/are used?

(b) What is the actual species that attacks the methylbenzene molecule?

(c) What *sort* of reagent is this?

(d) Write equations to show how it is formed in the reaction mixture.

CONTINUED

(e) There are three possible *mono*nitrated methylbenzenes. Draw them.

(f) Which one(s) is/are formed?

(g) Draw the mechanism of the reaction.

(h) Further nitration, to the di- and tri- substituted compounds, is more difficult. Why?

3. Similarly the first step in manufacture of TABT is by the nitration of 1,3,5-trichlorobenzene (fig 4).

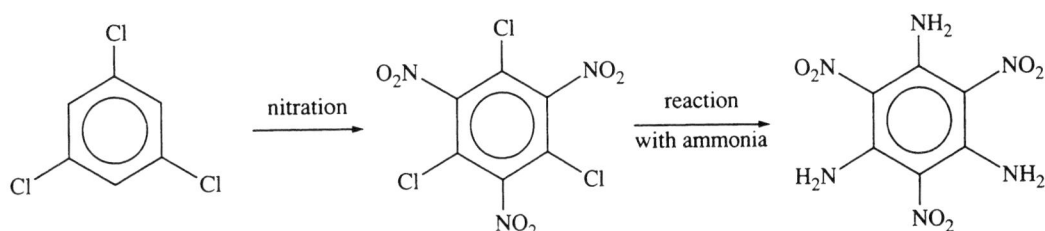

FIG 4

(a) Would you expect this to be easier or harder than the nitration of methylbenzene? Why?

(b) The second step involves attack by ammonia. But ammonia is a nucleophile, and aromatic chloro compounds are not easily attacked by nucleophiles. Why not?

(c) The presence of the three nitro groups (and the three chlorine atoms) makes this nucleophilic attack easier. Can you explain why?

4. A possible reaction scheme to make lead styphnate is shown in fig 5.

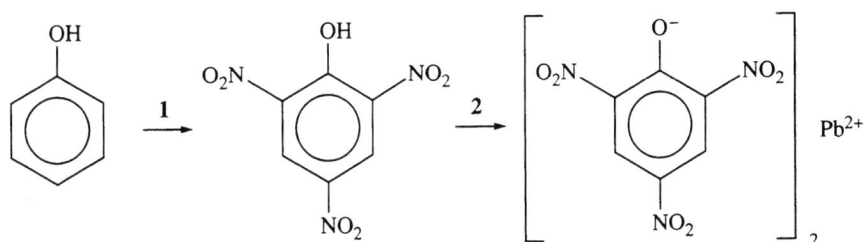

FIG 5

(a) Would (1) be easier than the nitration of methylbenzene?

(b) Explain your answer to (a).

(c) (2) might be carried out by reacting the 2,4,6-trinitrophenol (picric acid) with lead oxide or hydroxide. Write an equation for this reaction.

(d) It has been suggested that (2) could alternatively be done by reacting the picric acid with lead carbonate, but phenols don't react with carbonates. Why not?

(e) In fact, lead carbonate *will* work in this case. The properties of the phenolic $-OH$ group are modified by the presence of the three $-NO_2$ groups. How? What will their effect be and why?

CONTINUED

5. A synthesis of diazodinitrophenol – a detonator – has been suggested, starting from picric acid (fig 6).

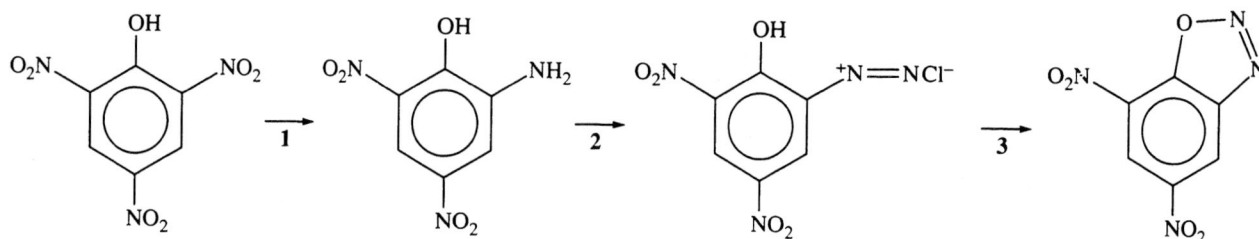

FIG 6

(a) Suggest a suitable reagent for (1).

(b) How could you prevent all the nitro groups from being reduced?

(c) Suggest suitable reagents and conditions for (2).

(d) Can you suggest a mechanism for (3)? (*Hint*: The −OH group has a lone pair of electrons.)

Sand castles and mud huts © 1991 Jeffrey Hancock, published by Hodder and Stoughton Educational

27. Spare parts

Like all machinery, our bodies wear out. And just as with machinery, a spare part may be all that is needed to repair it. These replacement parts range from the trivial and familiar, like fillings in teeth, to the amazing, such as heart valves, or even a new heart. The one feature they all share is the major use they make of synthetic polymers.

To be suitable for use in the body, the polymer must fulfil some stringent conditions. It should have no effect on the surrounding tissues, nor interfere with the normal processes of healing. No material yet found meets these requirements; even the most inert causes callouses to form around it. Several polymers are acceptable, however. More difficult still is blood compatibility; the clotting mechanism is so easily triggered that every artificial material sets it off.

If the polymer chains are of reasonably similar lengths and packed in an orderly fashion, the polymer will be solid. It may even be almost as regular as a crystal. For example, polytetrafluoroethene (Teflon or PTFE) is over 90% crystalline (fig 1). Other polymers may be much less regular, yet still solid at body temperatures, perhaps because their backbone is less flexible (Terylene), or because they are held into a glass-like lattice by dipole interactions (Terylene) or hydrogen bonding (nylon), or if there is extensive cross-linking between the chains. If the monomer units are particularly awkwardly shaped and non-polar, the polymer will be rubbery. The classic example of this is poly(dimethylsiloxane).

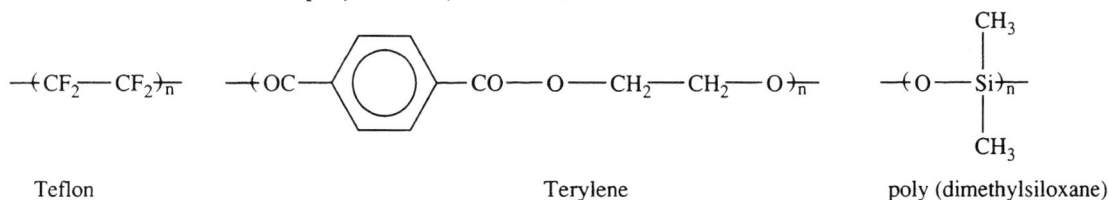

$$\begin{array}{lll}
-\!(CF_2\!-\!CF_2)_{\overline{n}} & -\!(OC\!-\!\bigcirc\!-\!CO\!-\!O\!-\!CH_2\!-\!CH_2\!-\!O\,)_{\overline{n}} & -\!(O\!-\!\underset{CH_3}{\overset{CH_3}{Si}}\!)_{\overline{n}}
\end{array}$$

Teflon   Terylene   poly (dimethylsiloxane)

FIG 1

Almost all hard body tissues can now be routinely repaired and operations such as hip replacements have become commonplace. Polymers are not strong enough to replace load-bearing bones, so plates or pins of stainless steel or titanium are used, often fixed in place by a glue formed by polymerising a monomer on to the join.

Questions

On the properties of polymers

1. A new polymer has been invented to repair ear drums as a possible cure for deafness. Before it can be tried out on human volunteers, it is tested on animals. Testing anything on animals is a controversial business.

(a) Briefly state two arguments in favour of it.

(b) Now give two arguments against it.

(c) Do you think that the polymer *should* be tested on animals?

(d) How would your answers to (a)–(c) be altered if it was a shampoo that was being tested?

CONTINUED

2. Explain:

    (a) what is meant by 'polytetrafluoroethene . . . is over 90% crystalline',

    (b) the difference between a crystalline and a glass-like lattice,

    (c) why cross-linking makes a polymer solid,

    (d) why 'if the monomer units are particularly awkwardly shaped and non-polar, the polymer will be rubbery'.

An early stimulus to the use of plastics in the body was provided just after World War II. A pilot who had survived a crash was found to have fragments of the aircraft's windscreen in his eye. The tiny pieces of Perspex had neither decomposed nor reacted with the eyeball. Perspex was an obvious first choice for contact lenses.

Two problems were immediately apparent. Perspex is a hard, rigid polymer, and the lenses were not always comfortable. More seriously, oxygen cannot pass through them. The cornea needs an oxygen supply for respiration ($2\text{--}5\ \text{mm}^3$ of O_2 per hour per cm^2 of cornea). If the oxygen supply is inadequate, the cornea respires anaerobically, producing 2-hydroxypropanoic acid, which makes the cornea swell and go hazy. Hard lenses containing Perspex are still used but they now also contain siloxanes, which allow more oxygen to permeate through.

A major development has been soft lenses. These are made of poly-HEMA (poly-(2-hydroxyethylmethacrylate)) usually copolymerised with NVP (poly-(N-vinylpyrrolidine) see fig 2).

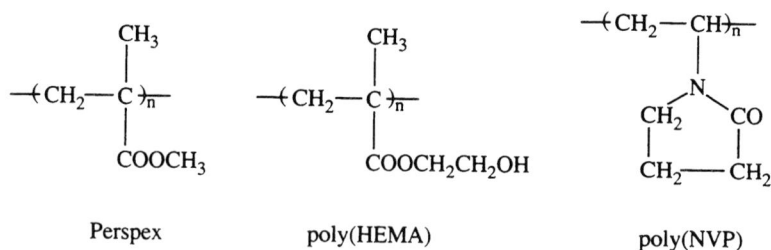

$$-(CH_2-\underset{\underset{COOCH_3}{|}}{\overset{\overset{CH_3}{|}}{C}})_n- \qquad -(CH_2-\underset{\underset{COOCH_2CH_2OH}{|}}{\overset{\overset{CH_3}{|}}{C}})_n-$$

poly(NVP):
$$-(CH_2-CH)_n-$$
with N attached to a ring: N bonded to CH_2 and CO; CH_2-CO at top, CH_2-CH_2 at bottom forming the pyrrolidine ring.

Perspex   poly(HEMA)   poly(NVP)

FIG 2

These polymers can absorb a large amount of water, changing from a hard plastic to a jelly. So they are very comfortable to wear.

As they are tough and strong, Perspex and related materials are also used in white dental fillings and false teeth. Pure Perspex shrinks a little when it is polymerised, so the fillings would fall out. Epoxy resins don't shrink, but they are difficult to polymerise in the tooth. (They are similar to those glues where you have to mix the contents of two different tubes.) Eventually a combination of the two materials was devised, to be fast setting without shrinking.

Ears, like eyes, have been repaired with polymers. The little bones of the middle ear that transmit the sound to the inner ear and thence to the brain can be replaced by pieces of Teflon, with an 85% chance of a hearing improvement. Hard, crystalline and inert, Teflon is also revolutionising blood vessel replacement. This is especially tricky because every synthetic material tends to cause the blood to clot. Gore-Tex appears to offer an elegant solution. This is a microgranular Teflon, with small pores between the granules. (It is also used in foul weather clothing, where the small pores allow the molecules of water vapour to pass through, and thus sweat to evaporate, but droplets of liquid water cannot enter from outside.) In artificial blood vessels, the pores allow the body tissue to colonise the Gore-Tex, so that eventually the blood is in contact with biological tissue rather than the plastic.

There is no limit to the soft tissues that can be rebuilt. The bones of

CONTINUED

Sand castles and mud huts © 1991 Jeffrey Hancock, published by Hodder and Stoughton Educational

the skull and face have been remodelled with Perspex, polyethene and polypropene, covered by a Terylene mesh with a graft of the patient's own skin on top. The skin grows into the plastic mesh and the repair can only be spotted by close examination. A breast can be remodelled by siloxanes in a silicone bag. Inserted under the skin of the chest, it is anchored in place as the body tissue grows into the Terylene mesh on the back of the bag.

But there is a problem to all this. None of these polymers is biodegradable. (That is why we can use them!) So when their owner dies, what happens? If he or she is buried, eventually the body decomposes – except for the artificial bits. What on earth will a future archaeologist think?

Questions

3. One method of polymerising alkenes is by a free radical addition reaction. This may be initiated by addition of a source of free radicals, such as dibenzoyl peroxide:

$$C_6H_5-\overset{\overset{O}{\|}}{C}-O-O-\overset{\overset{O}{\|}}{C}-C_6H_5 \rightarrow 2C_6H_5-\overset{\overset{O}{\|}}{C}-O\cdot$$

(a) Suggest how you could encourage this reaction to occur.

(b) The benzoyloxy radicals now react with an alkene, such as MMA (methylmethacrylate, more correctly called methyl 2-methylpropenoate):

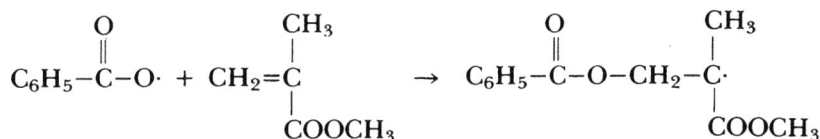

$$C_6H_5-\overset{\overset{O}{\|}}{C}-O\cdot \ + \ CH_2{=}\underset{\underset{COOCH_3}{|}}{\overset{\overset{CH_3}{|}}{C}} \quad \rightarrow \quad C_6H_5-\overset{\overset{O}{\|}}{C}-O-CH_2-\underset{\underset{COOCH_3}{|}}{\overset{\overset{CH_3}{|}}{C}}\cdot$$

This initiates a chain reaction. Why? Show what happens next.

(c) What stops ('terminates') the chain reaction?

(d) Dental filling material contains only small amounts of initiator, because too much of it leads to short polymer chains and a weaker material. Explain why.

(e) Alternatively, the polymerisation may be initiated by ultraviolet light. The dentist shines a small UV lamp at the filling for half a minute or so. How does this work?

4. Soft contact lenses are made of a copolymerised HEMA and NVP.

(a) Why does this material absorb so much water?

(b) Which constituent contributes most to this water-absorbing effect, the HEMA or the NVP? Explain.

(c) Why does this copolymer allow plenty of oxygen to permeate through to the cornea?

5. Terylene (called Dacron in the USA) is a polyester.

(a) How is it made? Give an equation and any special conditions.

(b) My lab coat is a terylene-cotton mixture. Why is it going into holes where I have spilt acids or alkalis on to it?

(c) Terylene is inert in the body, resisting decomposition well. Why?

CONTINUED

6. Siloxane polymers are made industrially from silicon and chloromethane. This might be represented schematically as follows (these aren't balanced equations):

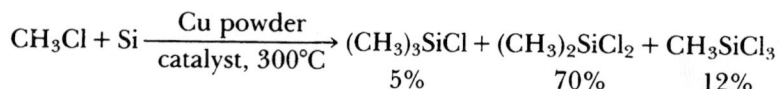

$$CH_3Cl + Si \xrightarrow[\text{catalyst, 300°C}]{\text{Cu powder}} \underset{5\%}{(CH_3)_3SiCl} + \underset{70\%}{(CH_3)_2SiCl_2} + \underset{12\%}{CH_3SiCl_3}$$

Dichlorodimethylsilane, for example, is then hydrolysed with water to make the diol, which then polymerises.

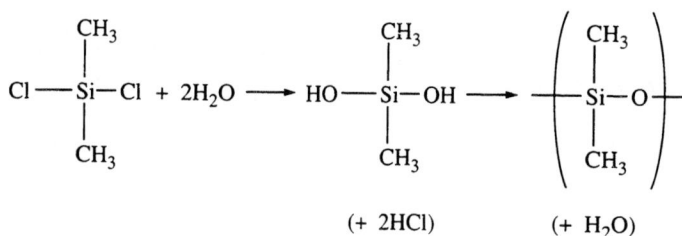

$$\underset{}{Cl-\overset{\overset{\textstyle CH_3}{|}}{\underset{\underset{\textstyle CH_3}{|}}{Si}}-Cl} + 2H_2O \longrightarrow \underset{(+\ 2HCl)}{HO-\overset{\overset{\textstyle CH_3}{|}}{\underset{\underset{\textstyle CH_3}{|}}{Si}}-OH} \longrightarrow \underset{(+\ H_2O)}{\left(\overset{\overset{\textstyle CH_3}{|}}{\underset{\underset{\textstyle CH_3}{|}}{Si}}-O\right)}$$

FIG 3

(a) The scheme in fig 3 shows what happens with dichlorodimethylsilane. What would be obtained if the same sequence of reactions was done with chlorotrimethylsilane ($(CH_3)_3SiCl$)?

(b) If there is more than a trace of the trimethyl compound in the mixture, smaller molecules of polymer are obtained with shorter chains. Why is this?

(c) What would happen if these reactions were done with trichloromethylsilane (CH_3SiCl_3)?

(d) If there is much of the trichloromethylsilane present, the polymer produced is very cross-linked. Explain this.

(e) About 100 000 breast implantations are done every year in the USA, 20% of them after surgery for cancer. The texture of the artificial breast must match that of the existing one as far as possible, whether soft or firm. How would you vary the starting materials for these different types of breast implant?

Sand castles and mud huts © 1991 *Jeffrey Hancock, published by Hodder and Stoughton Educational*

28. Headache or hangover?

Most painkillers, even the fizzy ones, contain one of three drugs; aspirin, paracetamol or ibuprofen.

Aspirin and its related compounds have been known for over two hundred years, ever since Edward Stone, the vicar of Chipping Norton, suggested that a tea made from the bark of willow trees might be good for 'aguish and intermittent fevers'. It was: it contained a compound of salicylic acid similar to aspirin. Aspirin itself was synthesised about a hundred years ago, closely followed by paracetamol. Ibuprofen is a more recent discovery; it was patented in 1962 (fig 1).

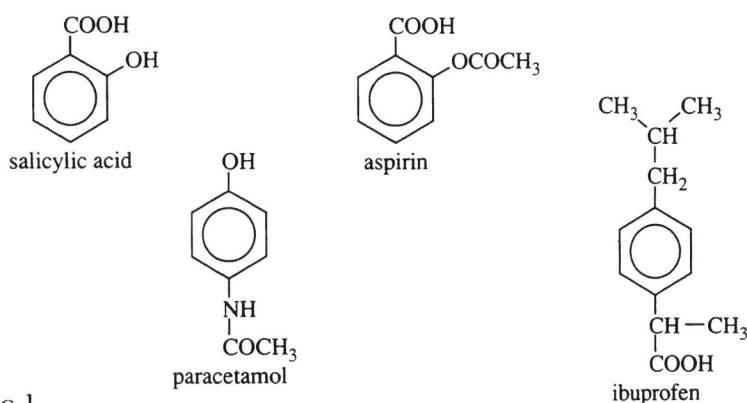

FIG 1

How do they work? Despite intensive research, nobody knows the full story, but ideas have recently started to appear.

There is a large group of compounds in the body called prostaglandins. They are implicated in a range of body processes, including headaches, inflammation and blood clotting (see Section 20: Of cholesterol and clots and corpses, p. 74). The formation of a blood clot is complicated, involving platelets – small cells present in the blood – and fibrinogen, a blood protein. When an injury occurs, the soluble fibrinogen changes into insoluble fibrin, which forms a network of fibres across the wound. The platelets aggregate (stick together in clumps) in this network, and so a clot builds up, sealing the wound. The aggregation of platelets is a vital step in this process and it seems to be controlled by two prostaglandins. One, thromboxane A_2, promotes aggregation. The other, prostacyclin, stops it. The correct tendency for the blood to clot is maintained by a balance between these two. Presumably the injury alters this balance, increasing the amounts of thromboxane A_2 to promote clotting.

Questions

1. The bark of the white willow tree contains salicylic acid, a white solid which is fairly soluble in hot water but insoluble in cold.

    (a) How would you name salicylic acid using the IUPAC system?

    (b) Briefly outline an experimental method you could use to obtain a sample of the solid acid from some willow bark.

2. (a) Aspirin is not very soluble in water (about $10 \, \text{g dm}^{-3}$ at body temperature). Suggest why (i) it is soluble at all, but (ii) not very soluble.

CONTINUED

(b) The calcium salt of aspirin is much more soluble (about $170\,g\,dm^{-3}$) and is sold as 'soluble aspirin' (fig 2). Why is it so much more soluble?

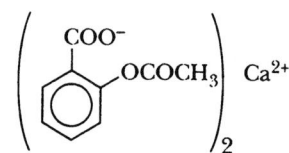

FIG 2

3. (a) Aspirin does not keep very well, especially in moist conditions. After some time, it starts to smell of vinegar. What is happening? Write an equation.

(b) Aqueous alkalis react even more rapidly with aspirin. Why? Write an equation for aspirin's reaction with aqueous sodium hydroxide.

(c) Calcium aspirin is made by reacting solid $CaCO_3$ with aspirin. Write an equation for this reaction.

(d) Why isn't aqueous $Ca(OH)_2$ used instead of solid $CaCO_3$?

(e) What happens to calcium aspirin when it arrives in the stomach, which contains dilute HCl?

4. Which of the four painkillers in fig 1:

(a) are phenols?

(b) is an ester?

(c) is a substituted amide?

(d) can exist as two optical isomers?

(e) would not give carbon dioxide if added to sodium carbonate solution?

Prostaglandins are synthesised in the body from arachidonic and γ-linolenic acids (fig 3). One of the early steps in the process is catalysed by cyclooxygenase, an enzyme found in many tissues of the body, including blood platelet cells. Cyclooxygenase from the platelets reacts with aspirin, and if radioactive aspirin is used (with tritium, ^3H, in the ethanoyl group), the cyclooxygenase becomes radioactive. The aspirin has apparently bound to the active site of the enzyme, and reacted with it to attach the CH_3CO- group to it (fig 4). (The hydrogen in **bold** is radioactive.) This ethanoylated cyclooxygenase has its active site blocked by the ethanoyl group, and so is unable to function. As a result, prostaglandin synthesis stops.

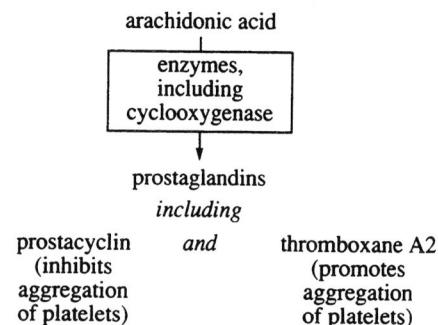

FIG 3

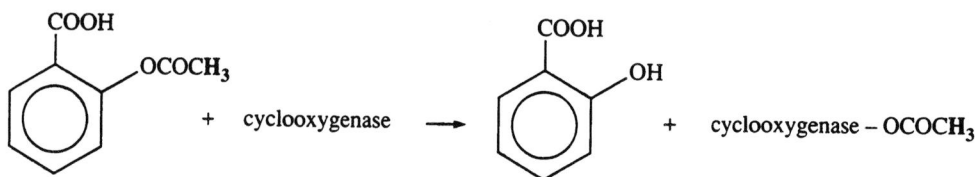

FIG 4

Can aspirin do this in the body? In an elegant experiment done in 1978, volunteers swallowed an aspirin tablet, and samples of their blood were taken at intervals. The platelets were separated out, then ^3H-labelled aspirin added. If the cyclooxygenase in the platelets had *already* been ethanoylated by the aspirin tablet, the enzyme couldn't react with the ^3H-labelled aspirin, so it wouldn't become radioactive. This was exactly what happened. Just one aspirin tablet was able to react with almost all the cyclooxygenase in the platelets. Other studies confirmed that aspirin stopped prostaglandin synthesis. One 40 mg tablet reduces thromboxane A_2 formation significantly, but prostacyclin formation is more resistant; 400 mg of aspirin are needed.

CONTINUED

This hundred-year-old drug now became the focus of great excitement, because blood clots in the wrong place are one cause of heart attacks. If aspirin can reduce the tendency of the blood to clot, can it reduce the likelihood of a heart attack? In 1988 this was shown to be true: patients who had already had one heart attack were less likely to have another if they were given 300 mg of aspirin a day.

How does this fit in with painkillers? Prostaglandins also cause headaches, fevers and inflammation, so it is likely that analgesics also work by interfering with prostaglandin synthesis, probably by ethanoylating an enzyme.

But the effect of aspirin on platelet cyclooxygenase lasts for several days, whereas aspirin works on your headache for only a few hours. And what about Mr Stone's willow bark tea? Salicylic acid and ibuprofen both reduce pain and fever, yet they cannot ethanoylate an enzyme as they have no ethanoyl group. Perhaps there are other mechanisms by which these drugs stop prostaglandin synthesis. Another hint about this has appeared recently.

An early step in prostaglandin synthesis involves the reaction of arachidonic acid with oxygen in a free radical reaction. (You may be as surprised as I am at the fact that free radical reactions can occur in the body at all. After all, they are notoriously difficult to control, and give rise to mixtures of products: think of the variety of compounds that results when methane and chlorine react.) Salicylic acid and many other anti-inflammatory drugs react with free radicals. If this occurs in the body, it presumably interferes with prostaglandin synthesis, and so may provide a second painkilling mechanism.

Questions

5. If taken in large doses, aspirin can cause stomach irritation, and even bleeding. (This was one of the great advantages of ibuprofen: it irritates the stomach to a much smaller extent.) It was originally thought that this was due to aspirin's acidic −COOH group. Why is this unlikely? (You might have two ideas.)

6. It has been suggested that the best dose of aspirin to reduce blood clotting is 40–50 mg. Explain how this figure is arrived at.

7. A possible laboratory synthesis of aspirin is given in fig 5, starting from methylbenzene.

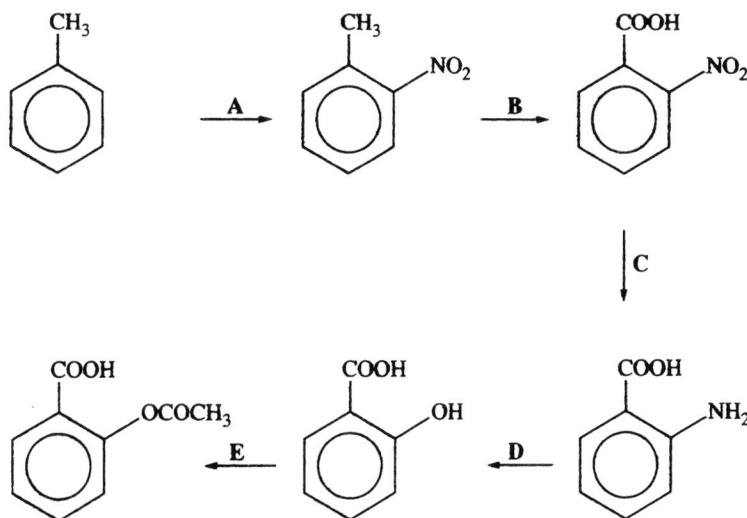

FIG 5

CONTINUED

(a) Identify A–E and specify any important conditions.

(b) Suggest how you would modify this synthesis to make the tritium labelled aspirin required for the experiments discussed in the text:

     (i) if you can start with any radioactive starting material you want,

     (ii) if only tritium labelled ethanol, $C^3H_3CH_2OH$, is available.

8. Aspirin is used now in large quantities, about 15 000 tonnes (15 000 Mg or 1.5×10^{10} g) a year in the USA alone. In industry the synthesis starts from benzene (the 'cumene process', fig 6).

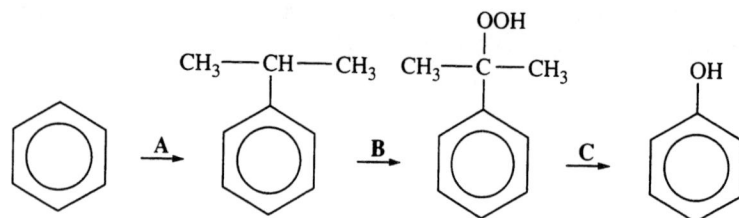

FIG 6

(a) Identify the reagents used in A–C.

(b) What is the other product of the process?

The phenol is now converted to salicylic acid by the reaction of the sodium salt with carbon dioxide at 125°C and 4–7 atmospheres pressure (fig 7).

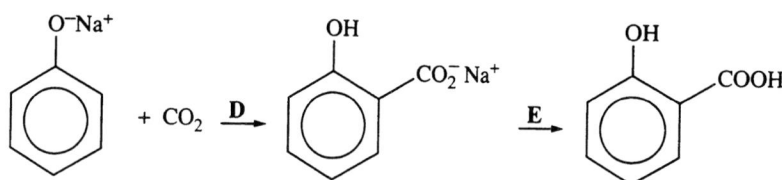

FIG 7

(c) What reagent is used in E to get the free salicylic acid from the sodium salt?

(d) Suggest a possible mechanism for D. (*Hint* The bonds in CO_2 will be polar.)

(e) Why is it better to use the sodium salt for D than phenol itself?

(f) After E there are two isomers present (fig 8). How do they both arise?

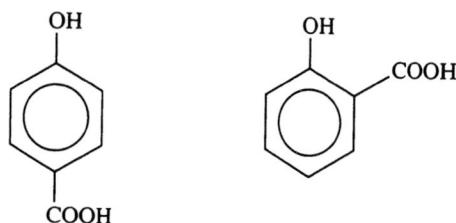

FIG 8

(g) They are separated by steam distillation. This works because the first isomer is much less volatile than the second. Suggest why this is so.

Sand castles and mud huts © 1991 Jeffrey Hancock, published by Hodder and Stoughton Educational

29. Impossible – yet perhaps it works?

'Painomin' is a commercially available analgesic. According to the manufacturer's specification, each tablet contains 500 mg paracetamol and 10 mg codeine. There is no aspirin in it.

Questions

1. (a) Design an experiment to show that paracetamol and codeine are present, and aspirin isn't. (The quantities of each drug aren't important.) You have access to chromatography apparatus and pure samples of aspirin, codeine and paracetamol.

(b) Chromatography doesn't enable you to be 100% certain that paracetamol and codeine were present. Why not?

(c) Could you alter your chromatographic method to increase your certainty?

2. In their advertising literature, the manufacturers claim that 'Painomin' is a more effective painkiller than either of its components on their own. You are to design an experiment to test this. You have access to 100 patients with post-operative pain in the surgical wards of a teaching hospital.

(a) There are several problems here that did not arise with the chromatography experiment. The first concerns the results. How can you assess whether a painkiller has worked and, if it has, whether it has worked better than another?

(b) Then there is the placebo effect. Whenever you give any group of patients a new drug, some of them will get better. Even if the new drug is only a sugar tablet, some of them will get better, *as long as they think they are taking a real drug*.

(i) Why do you suppose this is?

(ii) Can you plan the experiment to allow for the placebo effect?

(c) In modern drug trials the doctor does not know which drug she is giving to which patient. In this case, all the tablets would have to *look* identical.

(i) Why is it important that the doctor does not know which patient is receiving which drug?

(ii) How can this be organised?

(d) There are difficult *ethical* questions here. Some parts of the trial may not be to the benefit of the patient taking part. Explain why.

The standard drug trial design that is always used is the 'double blind' method. You may be able to see why it is called that.

CONTINUED

Homeopathy is a method of treatment of illness that has been in use for roughly two hundred years. It rests upon two principles:

Simillimum To be effective a medicine should produce the same symptoms as the illness. Very small doses of the medicine will then stimulate the body to combat the illness.
Succussion These small doses are prepared by diluting the original medicine, one drop into 99 drops of water (or ethanol/water mixture); at each dilution the mixture is 'succussed' (shaken violently). Without this succussion, the medicine is not effective.

This may seem odd. Odder still are the dilutions employed. Homeopathic practitioners speak of a dilution of 30c being quite usual, meaning that the hundred-fold dilution is repeated 30 times.

Questions

3. 1 cm^3 diluted to 1 dm^3 is a dilution of 1 in 10^3. Express a dilution of 30c in the same way: 1 in . . .?

4. Suppose that you started with a solution of the medicine which contained one mole of the substance in 1 dm^3. (Homeopathic medicines will not be a pure chemical, however, nor will they be used in such large volumes.)

   (a) How many molecules of the medicine are there per dm^3 in the original solution?

   (b) How many molecules are there per dm^3 of solution after it has been diluted to 30c?

Odd, isn't it? How can any solution so dilute actually *do* anything? Yet there are persistent reports of successful homeopathic treatments, and many doctors practise it.
A recent Glasgow study looked at over 100 patients with a history of hayfever; sneezing, runny nose and itching eyes. It is caused by pollen, and the homeopathic treatment is to give the patient a highly dilute suspension of pollen. The Glasgow team selected a dilution of 30c. The volunteers took one sugar tablet a day, to which the homeopathic solution had been added. Just 0.1 cm^3 of solution was added to a bottle of 28 tablets, so each tablet might have absorbed 0.1/28 cm^3 or 0.0036 cm^3 of this 30c solution.
How might they have designed the trial?

Question

5. Would you need to modify the design of your 'Painomin' trial to test the homeopathic tablets? In what way?

The results from the trial were clear. The homeopathic remedy caused a reduction in symptoms. The patients taking the homeopathic treatment were better. How can this be explained?
There may have been mistakes in the trial or even fraud. But the Glasgow team published a detailed account of their work, so that anyone can repeat and check the experiment.
Can you think of a theory to explain the Glasgow team's findings? How could a dilute solution interact with the body to set off the improvement? Just one molecule of a compound can trigger a response in some insects (see Section 21: Tonight, Josephine! p. 78), but what if there are *no* molecules of the drug present?

CONTINUED

One suggestion is that the original substance and the succussion cause specific water polymers to be formed, whose nature depends on the nature of the original drug. The shape and structure of these polymers then cause other polymers to be formed in the water as the original solution is diluted, so that they remain even after the original drug has been diluted out. These water polymers enable the solutions to affect the patient. Some evidence for this is said to have been obtained from spectroscopic studies. But could the polymers remain in solution for long enough? While the drug was diluted, and succussed, and stored, and measured out for the patient, and finally swallowed?

Question

6. So what should the scientific world do in the future?

(a) Decide that if a drug is pure water it can't have any effect. Any experiments which say otherwise are dishonest or mistaken. Stop funding any homeopathy trials and refuse to publish the results of any experiments.

(b) Carry on with experiments like the Glasgow study, but realise that a team of scientists running a trial like that will have less time for orthodox medicine.

(c) Carry on with proper trials, but insist that because homeopathy is so unlikely to work – and apparently contradicts conventional science – stronger proof is needed than for conventional drug trials.

What do you think? Support your case with reasons.

^{79}Br, chance $\frac{1}{2} \times \frac{1}{2} = \frac{1}{4}$. *Total* probability of one ^{79}Br and one ^{81}Br (RMM = 210) is $2 \times \frac{1}{4} = \frac{1}{2}$.

Chance that both bromines are ^{81}Br (total mass = 212) is $\frac{1}{2} \times \frac{1}{2} = \frac{1}{4}$. Thus probabilities (and therefore intensities) are 208:210:212 in the ratio of $\frac{1}{4}:\frac{1}{2}:\frac{1}{4} = 50:100:50$.

(e) Three peaks, masses 120, 122 and 124.

The peak at 120 arises from $CF_2{}^{35}Cl_2$, probability $\frac{3}{4} \times \frac{3}{4} = \frac{9}{16}$.

The peak at 122 arises from $CF_2{}^{35}Cl^{37}Cl$, probability $2 \times \frac{3}{4} \times \frac{1}{4} = \frac{6}{16}$.

The peak at 124 arises from $CF_2{}^{37}Cl_2$, probability $\frac{1}{4} \times \frac{1}{4} = \frac{1}{16}$.

Ratios = 9:6:1, so abundances = 100, 66.7 and 11.1, respectively.

(f) Three peaks, masses 164, 166 and 168. Ratios = 3:4:1.

Abundances = 75, 100 and 25 respectively.

5. Carbrital contains bromine (it is more usually known as carbromal). The prominent peaks at 208 and 210 and at 165 and 167 are of approximately equal abundances (like CF_3Br). Placidyl (more usually called ethchlorvynol) contains chlorine. Notice the peak at 115, about three times more abundant than the 117 peak (compare CF_3Cl).

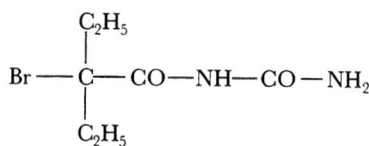

carbrital/carbromal

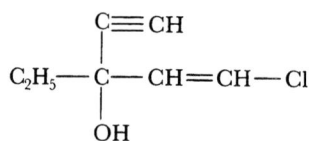

ethchorvynol

6.

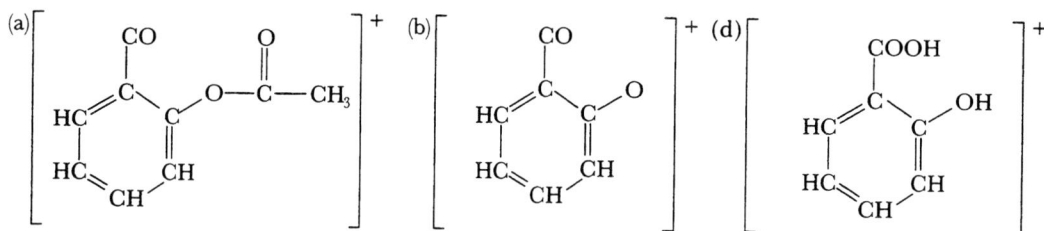

(c) $43 = [C_2H_3O]^+$.

7. (a) Three strong peaks occur in both spectra at 138, 120 and 92. There are other peaks common to both spectra.

(b)

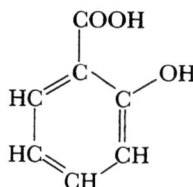

(c) $120 = [C_7H_4O_2]^+$; see 6(b) above.

(d) There is a small probability that one of the carbons in the molecule is ^{13}C, so the mass of M^+ will be greater by 1.

(e) But there is a negligible chance (see (f)) that two carbons in the same molecule are both ^{13}C, so a peak at $M^+ + 2$ is not detectable.

(f) There is a 1.1% chance that any carbon is ^{13}C, so with seven carbon atoms in the molecule there is a $7 \times 1.1\%$ chance = 7.7%

chance that *at least one* carbon will be ^{13}C, so the intensity is about 7.7% of the major peak.

4. Quantum theory and cancer

1. (a) Infra-red: electromagnetic radiation of longer wavelength than red light. Ultraviolet is beyond the violet end of the spectrum.
    (b) Raising of electrons to higher energy levels.
2. (a) Flame colour depends on the size of the electronic transition. Different elements have different energy levels, so the electronic transitions are of different sizes.
    (b) The Na emission spectrum is due to electrons falling from a higher energy level to a lower one. The absorption spectrum is caused by electrons doing the *same* transition in the opposite direction.
3. (a) (i) λ for midpoint ≃ 360 nm; frequency = 8.33×10^{14} Hz.
        (ii) energy = 332.6 kJ mol^{-1}.
    (b) (i) λ for midpoint ≃ 305 nm; frequency = 9.84×10^{14} Hz.
        (ii) energy = 392.6 kJ mol^{-1}.
    (c) UVB is of higher energy.
    (d) UVC is of higher energy still.
4. (a) More light than usual is emitted from the surface.
    (b) UV is of higher energy than visible: it is not possible to emit more energy than was absorbed. (First law of thermodynamics.)
5. (a) UVA rapidly converts the colourless melanin precursor into melanin.
    (b) Time has to be allowed for synthesis of more melanin precursor.
6. (a) More exposure to sunlight.
    (b) Melanin build-up (tanning) absorbs UVA, so reducing PLE.
    (c) Sunscreens protect against UVB, which are the main wavelengths causing soreness, so longer exposure without sunburn is possible. If the sunscreen has no UVA protection (and most don't), UVA exposure is increased by longer time in the sun.
7. (a) Sunscreen 1: 16.25, 17.0, 16.9; mean 16.7. Sunscreen 2: 11.25, 9.75, 11.25; mean 10.75. Perhaps thickness of application to skin varied.
    (b) (i) Sunscreen 1
        (ii) Sunscreen 2.
8. (a) Insoluble in water. Non-toxic. Colourless. Non-staining to clothes.
    (b) Benzene ring. C═O ring. Double bond on the atom next to the ring. Atom with non-bonding pair bonded to the ring.
    (c) Benzophenone-3: both UVA and UVB (although the cosmetic trade views this as protecting mainly against UVA). Octyldimethyl PABA: UVB. Octyl methoxycinnamate: UVB. Butyl methoxy dibenzoyl methane: UVA.
9. (a) Absorbance = 0.45 approx.
    (b) ε = 24900 mol^{-1} dm^3 cm^{-1}.
    (c) Absorbance = 5.15.
    (d) $I_0 = 1.4 \times 10^5$ I, $I = 7.1 \times 10^{-6} I_0$. Very little sun gets through! Of course, it will get wiped off, washed off if we swim, even absorbed by the skin.

5. Bitter-sweet

1. (a) Molecules that are able to bind with (receive) only certain (specific) taste molecules.
    (b) The tendency of an atom to attract electrons to itself within a covalent bond.

2. (a)

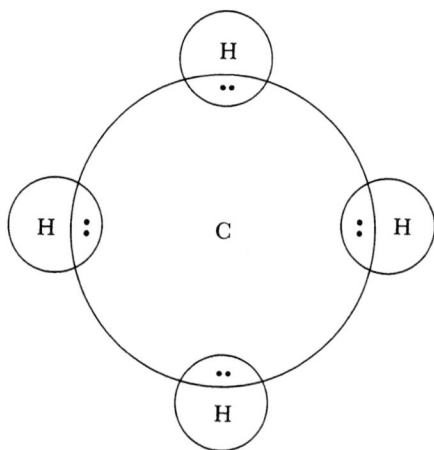

Shape: tetrahedral

(b) Non-bonding electron pair repels more and thus reduces the H—N—H bond angle.

(c) Non-linear, bond angle 104° (two non-bonding pairs).

3. (a) $\alpha \simeq 109°\,28'$, $\beta \simeq 104°$, $\gamma \simeq 109°\,28'$.

(b) C—H = 0.108; C—C = 0.154; C—O = 0.143; O—H = 0.096 nm.

(c) Scale drawing:

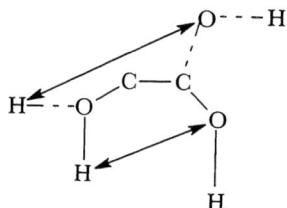

(d) AH—B distance can vary because of rotation about C—C and C—O bonds (as in the diagram above), between about 0.17 nm and 0.40 nm.

(e) Sketch:

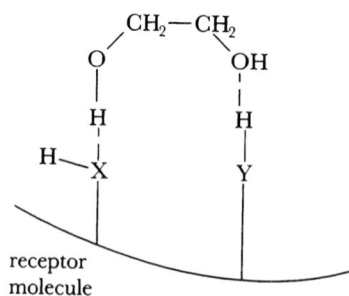

4. (a) Interaction between an H atom attached to a highly electronegative atom (N, O or F atom), and N, O or F atoms.

(b) (i) Unless the electronegativity of A is high, the $\delta+$ charge on the H is not high enough for a hydrogen bond to form.
(ii) Unless the electronegativity of X is high, the $\delta-$ charge on the X is not high enough for a hydrogen bond to form.

5. (a) No electronegative atoms; H-bonding not possible.

(b) Only one electronegative atom; formation of two H-bonds not possible.

(c) Two electronegative atoms, N and O, with H atoms at the right separation.

(d) Two electronegative atoms, but too far apart for H-bonding.

(e) Although the electronegativity of Cl is high (3.0), the Cl atom is too large for strong hydrogen bonding to be possible.

(f) Sulphur is not electronegative enough for H-bonding.

Sand castles and mud huts © 1991 Jeffrey Hancock, published by Hodder and Stoughton Educational

6.
(a)

Aspartame

(b)

Saccharin

(c)

Sodium cyclamate

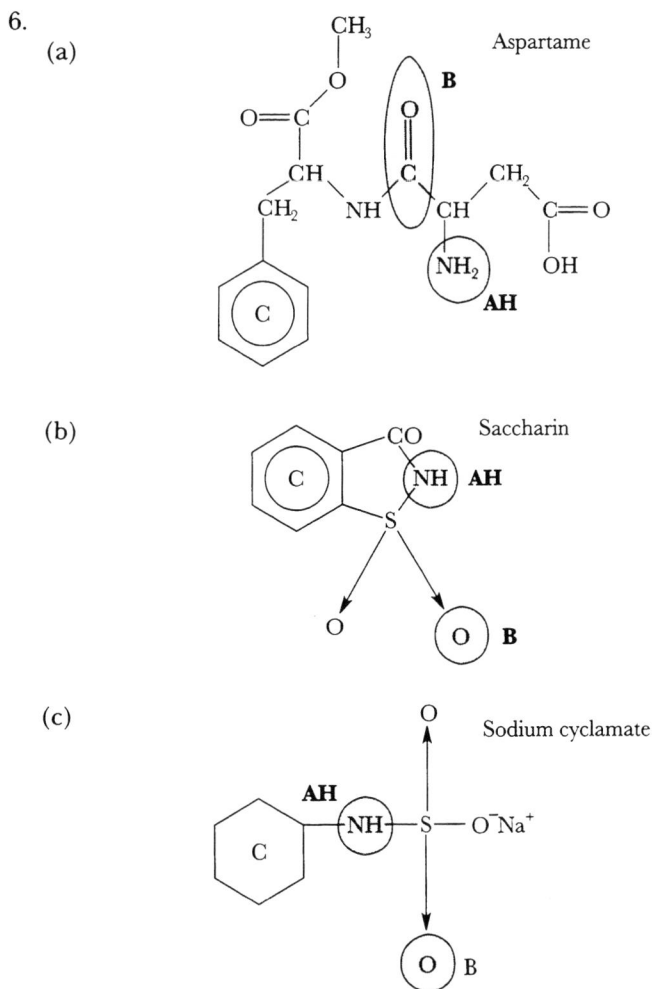

7. —NH—CO— group.
8. Molecule is too big to get into taste buds.
9. (a) 108°
   (b) 109° 28'
   (c) Planar arrangement involves bond angles of 108°; almost exactly right for saturated carbons.
   (d) Because the molecule is planar, the OH----O distance (z in fig 2) is less than 0.30 nm.

6. Anaesthetics

1. (a) N_2O is not a very strong anaesthetic; high doses are required.
   (b) Reversible reaction.
2. (a) Compounds bonded by electron transfer, resulting in positive and negative ions which attract each other.
   (b) Strong attractions hold the ions in a solid lattice.
   (c) Polar molecules have atoms of different electronegativities, so one end of the molecule acquires a small negative charge, the other end an equal positive one.

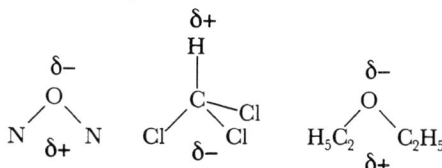

3. If a molecule is more blood-soluble, (a) it passes into the blood faster (and thence to the brain) and therefore acts faster, and (b) will be exhaled less readily.
4. (a) Xe. (Lowest anaesthetic pressure.)

(b) Anaesthetic pressure rises up the group; actual data are: He: >26400 kPa; Ne: 8920 kPa.

(c) High N_2 pressure required before N_2 acts as anaesthetic; achieved only at great depths.

(d) He has a much higher anaesthetic pressure, which isn't reached even at very great depths.

5. (a) Only polar molecules interact strongly enough with water molecules. This to compensate for loss of water–water hydrogen bonds.

(b) CF_4 is not polar, because it is symmetrical; CCl_2F_2 is polar.

(c) CCl_2F_2.

(d) (i) $CHClF_2$; more polar, so more water soluble, so a better anaesthetic.

(ii) CF_3—CH_3; ditto.

6. (a) Bonding: molecular covalent. Intermolecular forces: van der Waals'.

(b) Chance asymmetry in the electron distribution in one xenon atom leads to a temporary polarity in the atom. This in turn induces a polarity in an adjacent xenon atom, leading to attraction.

(c) Any non-polar liquid: benzene, any hydrocarbon, CCl_4, etc.

7. (a) Less polar; in fact, not polar at all, because it's symmetrical.

(b) CF_4.

(c) (i) $CHCl_3$; less polar, so more lipid soluble.

(ii) CF_3—$CHClBr$; ditto.

(d) CF_4 the better anaesthetic. 5(c) and 7(b) are directly opposite, so lipid solubility seems to be the more important factor.

(e) (i) $CHCl_3$, again because we must assume that lipid solubility is what matters most.

(ii) CF_3—$CHClBr$; ditto.

8. (a) Substances are often poisonous only because they react with a body chemical (e.g. CN^- with Fe^{3+}: see section 17, Life's little ironies).

(b) But some of these unreactive compounds did cause bad side-effects (convulsions, bad effects on blood pressure, etc.).

9. Gradient close to -1. The anaesthetic pressure is inversely proportional to the solubility of the compound in lipid. And since the effectiveness of an anaesthetic is inversely proportional to its pressure, this means:

Effectiveness *is proportional to* lipid solubility.

7. *Sand castles and mud huts*

1. (a) Oxygen: $1s^2\, 2s^2\, 2p^4$; silicon: $1s^2\, 2s^2\, 2p^6\, 3s^2\, 3p^2$.

(b) Covalent: shared pair of electrons.

(c) Oxygen has the orbitals available to accept only 2 more electrons; silicon can accept 4 (or more; see question 3b).

(d) Extended lattice: a solid structure held together by covalent bonds extending throughout the structure.

(e) Si can accommodate 8 (or more) electrons; at the edges they form three bonds, so have only 7 electrons.

2. (a) Hydrogen and oxygen have different electronegativities, so the O—H bond is polar. The molecule is non-linear, so the bond polarities don't cancel out.

(b) Hydrogen bonding in ice: attraction between the $\delta+$ H atom and the $\delta-$ O atom. This attraction is preferentially along the direction of the non-bonding pairs of electrons on the oxygen atom, so each water is surrounded tetrahedrally by 4 water molecules.

3. (a) A dative or coordinate bond.

(b) Before the formation of this bond, the silicon has 17 electrons

(14 of its own, plus 3 by formation of 3 covalent bonds to the oxygens). This means that there are 7 electrons in the outer third quantum shell. After the formation of the bond, the silicon has 19 electrons, 9 in the outer shell. This is possible only by octet expansion, using the low-lying 3d orbitals.

4. (a)

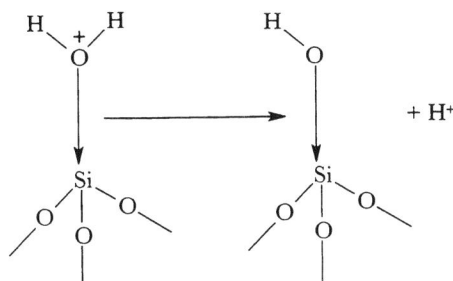

(b) Smaller particles have a bigger surface area, so more H^+ produced for a given amount of sand.

5. (a) The sand particles in a wet beach are virtually close packed.
   (b) Since the sand is (almost) close packed, it cannot get any closer as the small amount of water in it evaporates.

6. (a) Clay is not close packed, but has a relatively open structure with the spaces filled with water, so under stress the molecules can slide into the spaces.
   (b) Between flat microscope slides there are many hydrogen bonds, and to pull the slides apart, you have to break *all* the hydrogen bonds *at the same time*. If you slide them, you only break a few bonds at a time.

7. (a) Water loss.
   (b) When eventually the clay particles are more or less touching, the water left is structural water,
   (c) which binds the clay together by hydrogen bonding.
   (d) Even "dry" clay has water in it, and this holds the clay together by extensive hydrogen bonding.

8. (a) The jelly is formed by a coherent hydrogen bonded system. When tapped sharply, this system breaks down, making the mixture runny. On standing the hydrogen bonding reforms.
   (b) The paint is solid in the can, becomes runny under stress when brushed out, but resolidifies on the wall. (It will then dry normally.)

8. *Moles and things*

1. 1 mole of seconds $= 1.9 \times 10^{16}$ years; only about 10^{-6} mole of seconds since the beginning of time!

2. (a) $4CH_3NHNH_2 + 5N_2O_4 \rightarrow 9N_2 + 12H_2O + 4CO_2$
   (b) 2.4 tonnes.

3. (a) $10Al + 6NH_4ClO_4 \rightarrow 3N_2 + 9H_2O + 6HCl + 5Al_2O_3$
   (b) Mass of $NH_4ClO_4 = 418$ tonnes; mass of $HCl = 130$ tonnes.

4. If mass of washing up bowl is assumed to be M grams, volume $= 0.86 \times M \, dm^3$.

5. (a) About 1.1×10^{23} atoms.
   (b) About 1.7×10^{23} atoms.
   (c) First method. You cannot pack spheres perfectly; second method ignores this.

6. (a) Heat is absorbed in supplying latent heat until the hands are nearly dry.
   (b) 29.5 g.

7. (a) 756 g glucose.
   (b) 4.81 kg of water.

8. (a) Runner: about 30 moles of CO_2 or $720 \, dm^3$.
   (b) Petrol car: about 93 moles of CO_2 or $2240 \, dm^3$.

(c) Diesel car: about 90 moles of CO_2 or $2160\,dm^3$.

9. (a) This is 15.6 moles of CH_4, or $375\,dm^3$. This has interesting implications for vegetarianism and the greenhouse effect...

(b) Equivalent to $15.6 \times 30 = 468$ moles of CO_2.

10. About 1 molecule. (Insect releases about 8×10^{15} molecules. Volume of hemisphere is about $7 \times 10^{15}\,cm^3$.)

9. Import–export

1. (a) Control.
   (b) Closer to the plant. Probably in the direction of the prevailing wind.
   (c) High SO_2 levels damage foliage. (Compare SO_2 levels of 0.045 ppm and 0.008 ppm.)
   (d) Other effects – e.g. variation in rainfall, soil type, etc. – were not excluded.
   (e) Grow trees in greenhouse in identical conditions, with and without SO_2.

2. (a) 1.6947×10^6 tonnes.
   (b) $6.35 \times 10^8\,m^3$.
   (c) 2.59×10^6 tonnes.

3. (a) $1\,ppm = 1\,cm^3$ per m^3.
   (b) $1\,cm^3 = 1/24000$ moles $SO_2 = 64/24000\,g\ SO_2 = 2670\,\mu g\,m^{-3}$.

4. (a) Assuming 20 breaths per minute: $20 \times 500\,cm^3$ per $min = 10\,dm^3$. This gives $10 \times 60 \times 24 \times 366/10^3\,m^3$ per year $= 5270\,m^3$ of air. [If you forget that 1988 was actually a leap year, the figure is $5256\,m^3$ of air.]
   (b) 15.7 ppm means $15.7\,cm^3$ per m^3 of air $= 15.7 \times 5270 = 82750$ (82520) cm^3 of SO_2.
   (c) This is 82750/24000 (or 82520/24000) moles of SO_2 or H_2SO_3; about 3.45 (3.44) moles per year.

5. (a) 6400 tonnes SO_2.
   (b) $6400 \times 100/90 = 7111$ tonnes SO_2 produced.
   (c) From burning 3556 tonnes S.
   (d) % of S $= 3556 \times 100/1.2 \times 10^5 = 2.96\%$.
   (e) 17200 tonnes of $CaSO_4 \cdot 2H_2O$.
   (f) 1.2×10^5 tonnes C give 4.4×10^5 tonnes CO_2.
   (g) FGD adds 4400 tonnes each week; 1%. Trivial.

6. $571\,cm^3$.

7. $585\,cm^3$. (And since H_2S constitutes something like 0.03% of flatus, this suggests that each astronaut would produce about $1.96\,m^3$, or $1960\,dm^3$ a day! I don't know what NASA were thinking of!)

8. (a) 5.85×10^{-5} moles of acid and of SO_2.
   (b) $3.744 \times 10^{-3}\,g = 3744\,\mu g$ of SO_2.
   (c) (i) $1920\,\mu g\,m^{-3}$ (ii) 0.72 ppm.

9. Moor Lane, 1970: $954\,\mu g\,m^{-3}$
   Moor Lane, 1988: $100\,\mu g\,m^{-3}$
   Cox's Lane, 1988: $357\,\mu g\,m^{-3}$.
   (These are actual figures and show that although there has been a marked improvement in air quality, there is still some way to go, especially in industrial areas.)

10. Oxford, 1988: $195\,\mu g\,m^{-3}$.

10. Taking drugs

1. (a) (i) The constant of proportionality in the rate equation.
       (ii) The power to which the penicillin concentration is raised in the equation: i.e. n.
   (b) (i) Rate $= k[\text{penicillin}]^1$ (ii) Rate $= k[\text{penicillin}]^2$.

ANSWERS

2. (a)

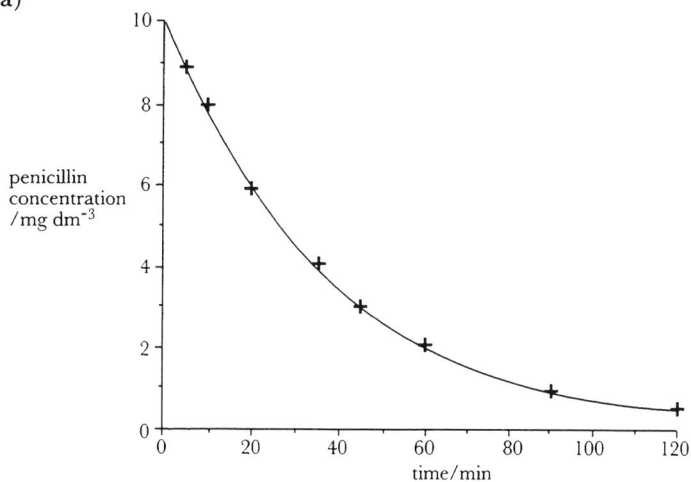

(b) The time taken for half the reagents to react.
(c) Close to 25 minutes in each case.
(d) Approximately 25 minutes. (Graph of log[penicillin] against time gives 25.8 min.)
(e) First. Only 1st order reactions have constant half life.
(f) Rate = k[penicillin]^1.
(g) min^{-1}.
(h) k = 2.8×10^{-3} min^{-1} (approximately).

3. (a) About 10.1 mg dm^{-3}.
(b) Volume = mass/concentration = 500/10.1 = 49.5 dm^3.

4. (a) (i) 1020/35 = 29.1 mg dm^{-3}
(ii) 29.1/2 = 14.6 mg dm^{-3}
(iii) 14.6/2 = 7.3 mg dm^{-3}
(iv) 7.3/2 = 3.6 mg dm^{-3}
(b) (i) 3.6 + 29.1 = 32.7 mg dm^{-3}
(ii) 32.7/2 = 16.4 mg dm^{-3}
(iii) 16.4/2 = 8.2 mg dm^{-3}
(iv) 8.2/2 = 4.1 mg dm^{-3}.

(c)

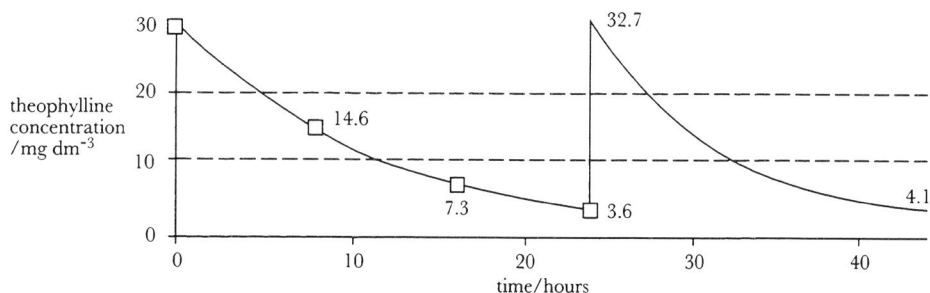

5. The same procedure:

Time, hours	0	8	16	24	32	40	48
Conc of drug/mg dm^{-3}	0	4.9	7.3	8.5	9.1	9.4	9.6
Conc after next dose	9.7	14.6	17.0	18.2	18.8	19.1	19.3

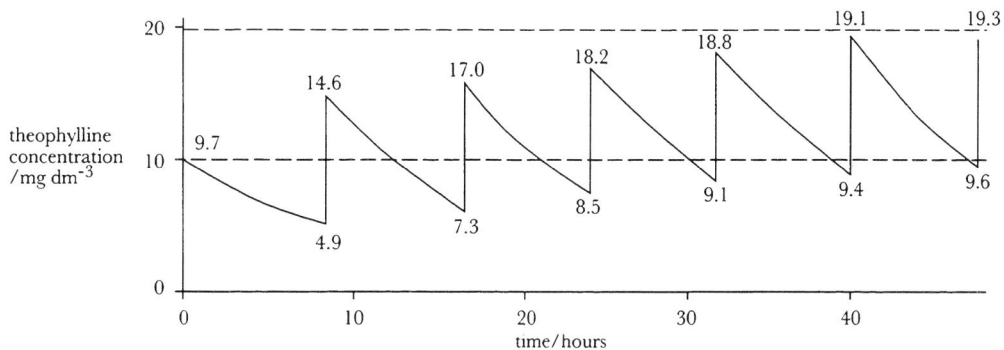

6. The dotted lines on the sketch graphs indicate the therapeutic and toxic doses. 340 mg every 8 hours falls within the boundaries much better, so this is the preferred dose.

7. Half lives! Penicillins have short one; chloroquine a half life of 9 ± 3 days.

8. (a) $16/42 = 0.38\,g/dm^3$ or $380\,mg/dm^3$ or $38\,mg/100\,cm^3$.
   (b) *About* 2 pints. (Of course, this depends on the distribution volume.)
   (c) Blood ethanol concentration $= 16/22 = 730\,mg/dm^3 = 73\,mg/100\,cm^3$ per pint, so she can't drink much more than a pint.

9. (a) About 11 hours (10.95 hours); around 9.30 am next morning.

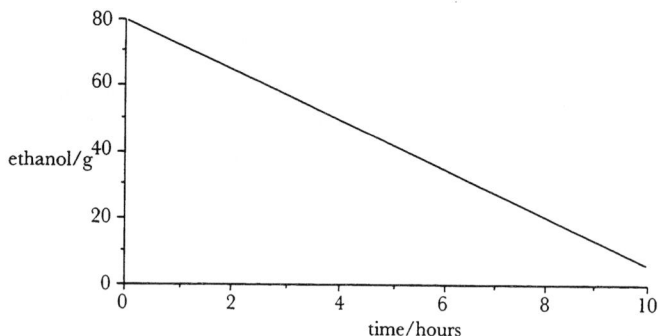

   (b) Zero. Rate $= k[\text{ethanol}]^0$, or Rate $= k$.
   (c) (i) 5.48 hours
       (ii) 2.74 hours
       (iii) 1.37 hours
       (iv) 0.68 hours.
   (d) Half life gets shorter. In fact, half life is proportional to initial concentration.

10. (a) $41\,mg/100\,cm^3 = 410\,mg/dm^3 = 0.410 \times 42\,g$ in his body $= 17.22\,g$.
    (b) Removed at $7.3\,g/hour$, so $4 \times 7.3\,g = 29.2\,g$ removed. Original total $= 17.22 + 29.2 = 46.42\,g$.
    (c) $110.5\,mg/100\,cm^3$ blood.
    (d) It is important to realise that the rate of breakdown of the ethanol ($7.3\,g$ per hour) is only an *average* rate for men. There is a large variation (see text). There will also be a range of distribution volumes.

11. A breakdown rate of 1 unit/hour $= 10\,cm^3$ ethanol/hour $= 7.9\,g$ ethanol/hour; close to the average value.

12. $80\,mg/100\,cm^3$ ethanol $= 0.017\,mol\,dm^{-3}$. $10\,mg\,dm^{-3}$ penicillin $= 3 \times 10^{-3}\,mol\,dm^{-3}$. Such a high ethanol concentration will swamp any enzymes, so the rate of breakdown of ethanol will depend only on the enzyme concentration.

11. *In the balance*

1. (a) $[CN^-] = 7 \times 10^{-6}\,mol\,dm^{-3}$.
   (b) $K = [Fe(CN)_6^{3-}]/[Fe^{3+}][CN^-]^6$; $K = 1.4 \times 10^{31}\,mol^{-6}\,dm^{18}$.

2. (a) $[R_3N]/[R_3NH^+] = 1.3 \times 10^{-7}$.
   (b) $[R_3N]/[R_3NH^+] = 0.16$
   (c) By injection. In the stomach, $[R_3NH^+]$ is too high; poor absorption.

3. (a) $K = [CuEDTA^{2-}]/[Cu^{2+}][EDTA^{4-}]$.
   (b) $[Cu^{2+}] = 9.4 \times 10^{-19}\,mol\,dm^{-3}$.
   (c) BAL removes *much* less Ca^{2+}.

4. (a) (i) $[Cl^-] = 4.5 \times 10^{-2}$;
       (ii) $[Cl^-] = 0.45$;
       (iii) $[Cl^-] = 4.5\,mol\,dm^{-3}$;
       (iv) $[Cl^-] = \infty$
   (b) No! − infinite concentration of reagent required.

5. (a) Shifts the equilibrium to the left.
   (b) This absorbs (most of) the added H^+.
   (c) Base reacts with the H^+; equilibrium then shifts to produce more H^+ which (almost) replaces the H^+ that has reacted.
   (d) Exhaling excess CO_2 lowers $[H_2CO_3]$, thus shifting the first equilibrium to the left and lowering the H^+ concentration. Thus the pH rises.
6. Increase it.
7. (a)
$$K_p = \frac{p^2CO_2\,pN_2}{p^2NO\,p^2CO}$$

   (b) At equilibrium, pressures of CO_2 and N_2 far exceed those of CO and NO.
   (c) System is not at equilibrium. (Forward and back reactions are too slow, because the activation energies are too high.)
   (d) Platinum gauze is a catalyst; system now rapidly goes to equilibrium.

12. Tooth decay

1. (a) $pH = -\log_{10}[H^+]$, where $[H^+] =$ concentration of H^+ ion in $mol\,dm^{-3}$.
   (b) Press indicator paper onto tooth; match colour to standards. Use glass electrode. [Actual experiments use volunteers needing crowns. Before that is done, glass microelectrodes implanted and experiments done. Then electrode removed and crowns made in the usual way.]
2. (a) Yes; as sugar consumption falls, so does decay, and vice versa.
   (b) Some other factor might have been at work.
   (c) Maybe the general health of children fell during the war, and tooth decay increased because of that? [Not really!]
3. (a) $K_a = \dfrac{[CH_3CH(OH)COO^-][H^+]}{[CH_3CH(OH)COOH]}$

   (b) $[H^+] = 2.0 \times 10^{-4}\,mol\,dm^{-3}$.
   (c) $[CH_3CH(OH)COOH] = 3.1 \times 10^{-4}\,mol\,dm^{-3}$.
   (d) Excess lactate ions shift equilibrium in 3(a) to the left, thus lowering the hydrogen ion concentration and raising the pH.
4. K_a for glycine $= 1.35 \times 10^{10}$, so [glycine] $= 296\,mol\,dm^{-3}$. Not possible!
5. (a) $K_1 = \dfrac{[CitH_2^-][H^+]}{[CitH_3]}$ $K_2 = \dfrac{[CitH^{2-}][H^+]}{[CitH_2^-]}$ $K_3 = \dfrac{[CitH^{3-}][H^+]}{[CitH^{2-}]}$

   (b) $K = \dfrac{[Cit^{3-}][H^+]^3}{[CitH_3]}$

   (c) $K_1 \times K_2 \times K_3 = \dfrac{[CitH_2^-][H^+]}{[CitH_3]} \times \dfrac{[CitH^{2-}][H^+]}{[CitH_2^-]} \times \dfrac{[CitH^{3-}][H^+]}{[CitH^{2-}]}$

   then cancel down. To find pK, take $-\log_{10}$ throughout.
   (d) $pK = 14.3$, so $K = 5.0 \times 10^{-15}\,mol^3\,dm^{-9}$.
   (e) If x moles per dm^3 of $CitH_3$ dissociate:
$$CitH_3 \rightleftharpoons Cit^{3-} + 3H^+$$
   $\qquad$ 0.1–x $\qquad$ x $\qquad$ 3x mol dm^{-3}.
   Then $K = (3x)^3 x / 0.1–x$. But $0.1–x \approx 0.1$, so
   $27x^4 = 0.1 \times 5.0 \times 10^{-15}$, and $x = 6.6 \times 10^{-5}\,mol\,dm^{-3}$. $pH = 4.2$.

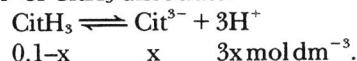

---

6. (a) $NH_3 + H^+ \rightarrow NH_4^+$.
  (b) A solution that will hold the pH closely constant even if small amounts of acid or base are added.
  (c) If acid is added, it reacts with the HCO_3^-, thus largely being absorbed. If base is added, it reacts with the H^+ ions, and the H_2CO_3 ionises to produce more H^+. Either way, the pH is little altered.
  (d) $2.5 \times 10^{-3} \, mol \, dm^{-3}$.
  (e) $PO_4^{3-} + H^+ \rightleftharpoons HPO_4^{2-}$; $HPO_4^{2-} + H^+ \rightleftharpoons H_2PO_4^-$; the second more important.

7. Chewing removes food debris, and stimulates flow of saliva, which returns pH to 7.0.

8. (a) Plaque bacteria would convert them to acids.
  (b) Ca^{2+} ions shift eq [1] to the left; Ca^{2+} and PO_4^{3-} ions both shift eq [2] to the left. This reforms tooth material.
  (c) pH 6.0 has higher $[H^+]$, thus pushing equilibrium to right, and lowering $[F^-]$. Thus pH 8.0 better.

9. (a) $5 \times 10^6 \, g \, H_2O$; i.e. $5 \, m^3$.
  (b) Removes freedom of choice from individual. Any drug treatment should be done only with the consent of every person receiving it. Not needed; people can take NaF tablets. Expensive.

10. Mean volume with no decay: $111 \, cm^3$. Mean volume with much decay: $99 \, cm^3$. Thus saliva of children with no decay was a more effective buffer, thus holding pH of plaque nearer to 7.0.

13. Blood, sweat and seas

1. (a) 6.6 lattice.
  (b)

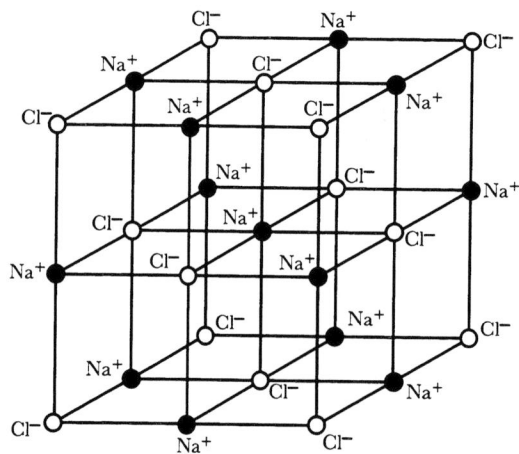

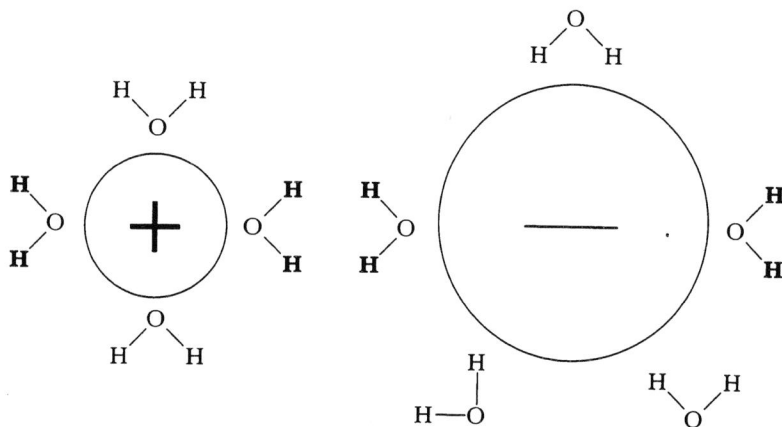

  (c) Hydrated: surrounded by water molecules, with the oxygen (the $\delta-$ end of the water dipole) pointed towards the cation, and vice versa for the anion.

2. Water has a much lower boiling point (and hence lower vapour pressure at ambient temperature) than NaCl, because of the weaker forces between the particles (hydrogen bonding as opposed to ionic interactions).

3. (a) Lattice enthalpy: enthalpy change per mole for
    $Na^+(g) + Cl^-(g) \rightarrow NaCl(s)$ (or vice versa, with opposite sign).
    Hydration enthalpy: enthalpy change per mole for
    $Na^+(g) + aq \rightarrow Na^+(aq)$
    Solution enthalpy: ΔH for $NaCl(s) + aq \rightarrow Na^+(aq) + Cl(aq)$.
  (b) Radius of $Na^+ <$ radius of K^+; stronger attraction for water dipole.

    (c) NaCl: $780 - 406 - 364 = +10\,kJ\,mol^{-1}$.
       KCl: $711 - 322 - 364 = +25\,kJ\,mol^{-1}$.

    (d) KCl less soluble because it is more endothermic, so less favourable.

    (e) Endothermic, so unfavourable.

    (f) But the entropy of the process is favourable, so overall ΔG is negative and so favourable.

4. (a) Calcium ions have double the charge of sodium ions.

    (b) Vast excess of sodium ions.

    (c) Na^+ smaller than K^+ because it has one less quantum shell of electrons. Ionic charge is thus more concentrated, so attraction stronger.

    (d) Li^+ has many more water molecules in several layers associated with it.

    (e) Smaller ions attracted most strongly. Smallest hydrated ion is K^+.

5. (a) Logic: [total number of atoms in ring]-crown-(number of oxygens).

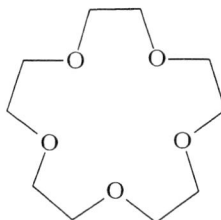

    (b) ΔH of solvation by CH_2Cl_2 insufficient to compensate for large lattice enthalpy of the ionic solid.

    (c) Very favourable interaction of cation with crown ether helps to compensate for loss of the lattice enthalpy.

    (d) (i) [14]-crown-4, LiCl is most soluble; [15]-crown-5, NaCl most soluble and so on.
       (ii) Effect of cation size: "hole" in [14]-crown-4 just right size for the oxygens to be able to interact well with Li^+, hole in [15]-crown-5 exactly right size for Na^+ ion and so on.

6. (a) Presumably also a size effect, as with the crown ethers.

    (b) Li^+ smaller than Na^+ or K^+, so any molecule(s) that interact with Na^+ or K^+ efficiently won't pick up Li^+ nearly so well.

14. Five per cent of us

1. (a) About 6.

    (b) It still has some solubility, even at this pH.

    (c) About $10^{-6}\,mol\,dm^{-3}$; i.e. $2.7 \times 10^{-5}\,g\,dm^{-3}$.

2. <0.2 ppm. "Later figures probably more reliable."

3. (a) Diagram of $[Al(H_2O)_6]^{3+}$ ion. (See, for example, the answer to question 1(d), section 13, above.)

    (b) $[Al(H_2O)_6]^{3+} + H_2O \rightleftharpoons [Al(H_2O)_5OH]^{2+} + H_3O^+$
       $[Al(H_2O)_5OH]^{2+} + H_2O \rightleftharpoons [Al(H_2O)_4(OH)_2]^+ + H_3O^+$

    (c) Because of the H_3O^+ produced above.

    (d) Lowering pH shifts equilibria to the left; raising it, to the right.

    (e) H_3O^+, produced as above, reacts with the HCO_3^- ions as follows:
       $H_3O^+ + HCO_3^- \rightarrow 2H_2O + CO_2$

4. (a) $[Al(H_2O)_4(OH)_2]^+ + OH^- \rightleftharpoons [Al(OH)_3(H_2O)_3]$
      $[Al(OH)_3(H_2O)_3] + OH^- \rightleftharpoons [Al(OH)_4(H_2O)_2]^-$ and even
      $[Al(OH)_4(H_2O)_2]^- + 2OH^- \rightleftharpoons [Al(OH)_6]^{3-}$

    (b) As pH is raised, they shift to the right.

5. (a) As pH is raised, less charged, and therefore less soluble ions
        are produced, as well as the insoluble $[Al(OH)_3(H_2O)_3]$.
        Above about pH 7, anionic species, soluble once more, are
        formed.
    (b) Stomach: $[Al(H_2O)_6]^{3+}$; soluble. Duodenum:
        $[Al(OH)_4(H_2O)_2]^-$; quite soluble, possibly less than in
        stomach.
    (c) Probably quite soluble in both, more in stomach. In the
        stomach:
        $AlO(OH) + 3H^+ + 4H_2O \rightarrow [Al(H_2O)_6]^{3+}$, while in the
        duodenum:
        $AlO(OH) + OH^- + 3H_2O \rightarrow [Al(OH)_4(H_2O)_2]^-$
6. Rhubarb and apples are both more acidic than potatoes.
7. (a) $Al + 3H^+ + 6H_2O \rightarrow [Al(H_2O)_6]^{3+} + 1\frac{1}{2}H_2$
        $Al + OH^- + 5H_2O \rightarrow [Al(OH)_4(H_2O)_2]^- + 1\frac{1}{2}H_2$
    (b) As the pH is raised the first of these reactions becomes less
        favourable. Eventually, above pH 7, the second equation
        increases in importance.
8. (a) $[AlF_6]^{3-}$
    (b) $[Al(H_2O)_6]^{3+} + 6F^- \rightleftharpoons [AlF_6]^{3-} + 6H_2O$
    (c) F^- ions are negatively charged; they therefore have a stronger
        attraction for Al^{3+} ions than neutral water molecules do.
    (d) Equilibrium in (b) pulls equilibria such as those in 7(a) to the
        right.
9. (a) $[Al(PO_4)_3]^{6-}$
    (b) $AlPO_4$ has no charge so interaction with water molecules is
        weak. (Same arguments for solubilities of, e.g.
        $[Al(OH)_3(H_2O)_3]$ compared to $[Al(H_2O)_6]^{3+}$.)
    (c) Phosphate ions are highly charged; the same argument as for
        F^- ions.
    (d) Phosphoric acid lowers the pH (see fig 2). Phosphate ions, like
        F^- ions, increases tendency of Al to dissolve in water. [On the
        other hand, amounts added are small.]
10. (a) The two ions have the same charge and similar ionic radius.
    (b) The ionic radii are not the same, so the Al^{3+} doesn't fit into
        the Fe^{3+} binding site very well.
    (c) Ionic radius of Ga^{3+} is closer to that of Fe^{3+}.

15. Rocket fuels

1. (a) A compound whose formation is endothermic; i.e. $\Delta H_f^{\ominus}$ is
        positive.
    (b) Distillation: boil the water-hydrazine hydrate mixture and
        condense the vapour coming off at 118°C.
2. (a) −3   (b) −2   (c) −1   (d) 0
    (e) −3   (f) +4
3. (a) $50.6 \, kJ \, mol^{-1}$.
    (b) The activation energy for the decomposition is too high.
    (c) Heat it. This speeds up the reaction because more molecules
        now have energy greater than the activation energy.
    (d) The metal gauze acts as a catalyst, lowering the activation
        energy and thus allowing rapid reaction.
4. (a)   (b)

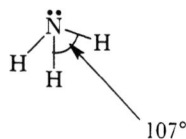

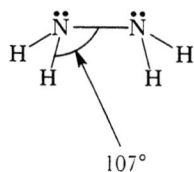

    (c) $NH_3 + H_2O \rightleftharpoons NH_4^+ + OH^-$
    (d) Non-bonding pair of electrons.

(e) Yes, because it also has non-bonding pairs of electrons.
$N_2H_4 + H_2O \rightleftharpoons N_2H_5^+ + OH^-$ (Formation of $N_2H_6^{2+}$ *very* unlikely.)

5. (a) N_2H_4 is polar (N is more electronegative than H; unlike C_2H_4, N_2H_4 is not symmetrical), so it has quite strong dipole–dipole interactions. C_2H_4 has only van der Waals' forces.
   (b) N_2H_4 is polar (see above), and it can form $N_2H_5^+$ which being charged will be very soluble.
   (c) The methyl groups make for poorer packing in the solid, so weaker inter-molecular forces.

6. (a) $N\equiv N$ shorter than $C\equiv C$ (greater nuclear charge in N).
   (b) N—N, O—O and F—F are all short bonds, short enough for strong repulsions between non-bonding pairs of electrons on adjacent atoms.
   (c) $-95\,kJ\,mol^{-1}$.
   (d) Weak N—N bonds and strong $N\equiv N$ bonds.

7. (a) $-538.8\,kJ\,mol^{-1}$
   (b) N_2O_4 is endothermic to the extent of $9.2\,kJ\,mol^{-1}$, so when N_2O_4 react with 2 hydrazines, $9.2/2 = 4.6\,kJ$ extra energy available per mole of N_2H_4.

8. (a) Carbon is $\delta+$, oxygen $\delta-$ because of the electronegativity difference between C and O.
   (b) Nucleophilic attack by: NH_2— on the $C{=}O$ group.
   (c) The same!

16. Most dangerous, most vital

1. (a) HF is *extremely* polar (F is small and very electronegative), so strong H-bonds formed between molecules.
   (b) F has three non-bonding pairs of electrons which reduce α to close to the tetrahedral angle (109° 28').
   (c) NH_3 (one non-bonding pair) has bond angle 107°. OH_2 (two non-bonding pairs) has bond angle about 104.5°, so with three non-bonding pairs the HF angle should be less than this. In fact, thermal motions stretch the chain, thus increasing α to about 120°.
   (d)

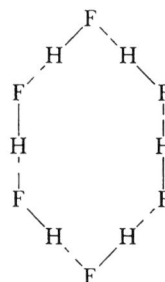

2. (a)

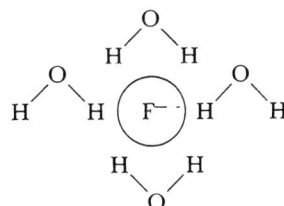

   (b) F^-----H—F
   (Actually the H-bond is so strong that the ion is symmetrical.)
   (c) The F^- ions interact more strongly with HF than H_2O because HF is so polar.

3. (a) HF is very polar.
   (b) 4HF molecules around one water molecule, hydrogen bonded as in ice: two HF through the H to the lone pair of the O, two through the F to the H atoms of the water.

4. (a) F^- is a very small anion, so fluorides have a very high lattice energy.

(b) Insoluble fluorides, either MgF_2 or CaF_2 formed, thus removing F^- ions from solution in the tissues.

(c) This depends on the position of the equilibrium:
$2Ca_5(PO_4)_3X(s) + 2H^+(aq) \rightleftharpoons 3Ca_3(PO_4)_2(s) + Ca^{2+}(aq) + 2HX$, where X can be OH^- or F^- ions. The position of the equilibrium will depend on the affinity of X for the H^+ or the solid lattice. F^- is smaller than OH^-, thus making the fluoride's solid lattice more stable. In addition, F^- is a weaker base than OH^-, thus having a lower affinity for H^+ ions.

5. (a) Donation of one pair of e from each F^-, making 6 pairs in all on the Al.

(b) 12. Al uses low-lying empty 3d orbitals.

(c) Octahedral.

(d) The Al—Cl bond is too weak to make it favourable to promote electrons to 3d orbitals. There may also be problems of the large size of Cl atoms, too.

(e) Boron has no low-lying empty orbitals (2d orbitals don't exist).

6. (a) $Sn + 2HF \rightarrow SnF_2 + H_2$; $SnO + 2HF \rightarrow SnF_2 + H_2O$.

(b) F_2 very strong oxidant; always tends to bring out the highest oxidation state of a metal. $Sn + 2F_2 \rightarrow SnF_4$.

7. There is no stronger oxidising agent than F_2, so chemical oxidation just isn't possible. Although there are stronger reducing agents that Al (even C if the temperature is high enough), these reductants or the temperatures required are too costly.

8. Strong C—C and C—F bonds. In addition, the F atoms are of just the right size to pack tightly round, and thus protect, the C—C backbone.

17. Life's little ironies

1. (a) Fe^{2+} and Fe^{3+}.

(b) Haemoglobin, methaemoglobin, myoglobin, cytochromes.

(c) Catalase is an iron-containing catalyst.

(d) Haemoglobin, carboxyhaemoglobin etc.

2. (a) $1s^2 2s^2 2p^6 3s^2 3p^6 3d^6 4s^2$.

(b) $1s^2 2s^2 2p^6 3s^2 3p^6 3d^6$.

(c) $1s^2 2s^2 2p^6 3s^2 3p^6 3d^5$.

3. (a) Atoms, molecules or ions linked to a central atom in a complex ion.

(b) Non-bonding pair of electrons.

(c) Donation of this non-bonding pair into the orbitals of the central ion.

(d) CH_4 has no non-bonding pair of electrons.

4. (a) The iron ions attract water molecules to form species such as $[Fe(H_2O)_6]^{2+}$ and $[Fe(H_2O)_6]^{3+}$.

(b) The cation polarises water molecules thus:
$$H_2O + [Fe(H_2O)_6]^{2+} \rightleftharpoons [Fe(H_2O)_5(OH)]^+ + H_3O^+$$
Fe^{3+} does this more than Fe^{2+}; higher charge (and smaller size) make it more polarising.

(c) Acid stomach suppresses hydrolysis; alkaline intestine increases it.

5. Paleness: less haemoglobin is present to colour the skin, and less O_2 can be transported, resulting in tiredness (less O_2 available for muscles).

6. (a) CN^- has a negative charge, so a much greater electron density. Electron pair donation is therefore easier/stronger.

(b) Fe^{3+} has a greater charge than Fe^{2+}, so attracts anions more strongly.

(c) Co^{2+} ions also form CN^- complexes, so less CN^- available to bind to Fe^{3+}.

(d) Amyl nitrite oxidises some of the haemoglobin to

methaemoglobin, which complexes strongly with the CN^- ions. Thus there are fewer free CN^- ions to attack the cytochromes.

7. (a) There is more haemoglobin, so more O_2 can be transported.
   (b) In the city, CO complexes with some of the haemoglobin, so less is available to transport O_2.
   (c) No. If there was less catalase present to destroy H_2O_2, the H_2O_2 would oxidise haemoglobin to methaemoglobin, which cannot transport O_2.

8. (a) $[Fe(CN)_6]^{4-}$ is more easily oxidised; less positive $E°$.
   (b) $E°$ for $[Hb-Fe^{3+}]$, $[Hb-Fe^{2+}]$ should be between $+1.23$ and $+1.77V$. (Physiological conditions won't be standard, however, so this may not be true.)

9. Increasing $E°$, so each is able to oxidise the preceding one: b, c, a.

10. (a) Non-linear; the O uses one non-bonding pair to bond to Fe, so one non-bonding pair is left. Thus angle is just less than $120°$. (In fact, the actual shape is not yet known with certainty, but it is probably this bent one.)
    (b) C has one non-bonding pair, used to bond to Fe, so linear.

18. Ozone problems

1. (a) Species with unpaired electron(s).
   (b) Unpaired electron(s). Possessing the energy used to homolyse the bond, so very reactive. Short lived.
   (c) Two unpaired electrons. *Extremely* reactive.

2. (a) $f = 497000/6.02 \times 10^{23} \times 6.63 \times 10^{-34} = 1.25 \times 10^{15}\,Hz$.
   (b) $\lambda = 3.00 \times 10^8/1.25 \times 10^{15} = 2.41 \times 10^{-7}\,m = 241\,nm$.
   (c) Ozone formation (reaction [2]) releases energy.

3. (a) $497\,kJ$.
   (b) $497\,kJ$.
   (c) Third body needed to absorb energy, or else the O_2 immediately homolyses again.

4. (a) $Cl_2 \rightarrow 2Cl\cdot$ then
   $CH_4 + Cl\cdot \rightarrow CH_3\cdot + HCl$, and $CH_3\cdot + Cl_2 \rightarrow CH_3Cl + Cl\cdot$ etc.
   (b) Product of one reaction step is a reagent for the other, and vice versa.
   (c) [6] and [7]. Product of [6] is a reagent for [7], and vice versa.
   (d) Reaction of two free radicals. In the atmosphere the concentration of radicals is so low, that this is *very* improbable.

5. (a) $NO + O_3 \rightarrow NO_2 + O_2$, then $NO_2 + \cdot O\cdot \rightarrow NO + O_2$
   (b) $\cdot N = O$. Both $Cl\cdot$ and $\cdot NO$ have an unpaired electron.

6. Surely the effect of plant growth is most to be feared? We depend on plants for food, O_2, etc.

7. (a) Increase in number of cars on the roads; rush hours.
   (b) Sunrise.
   (c) Presumably other reactions start to destroy the ozone.

8. (a) Ozone levels very low, so light absorption is only significant in a long light beam.
   (b) $280-290\,nm$ (see fig 1).
   (c) Variations in weather; some days cloudy: less O_3.
   (d) Coventry max $\approx 0.075\,ppm$; compare LA about $0.4-0.5\,ppm$.
   (e) More intense sunlight, more cars.

19. Fatty foods

1. (a) A group containing C and H atoms only.
   (b) Saturated means containing only $C-C$ single bonds. Unsaturated compounds contain double or triple carbon–carbon bonds.

(c) Further substitution. Reduced by use of excess NH_3.

(d) A, because of the problems of further substitution with NH_3.

(e) It would react more rapidly and give a better yield, because the C—I bond is weaker than C—Br bond.

4. (a) Double C=C bond between carbons 9 and 10.

(b) Elimination.

(c) KOH in ethanol, refluxed.

(d) $CH_3(CH_2)_6CH=CH(CH_2)_{13}CH_3$, i.e. 8-tricosene, cis and trans, as well as trans-9-tricosene and the secondary alcohol: $CH_3(CH_2)_7CH(OH)CH_2(CH_2)_{12}CH_3$.

(e) The two parts of the hydrocarbon chain on the same side of the C=C bond:

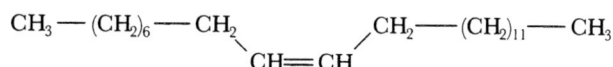

$$CH_3 - (CH_2)_6 - CH_2 \diagdown \qquad \diagup CH_2 - (CH_2)_{11} - CH_3$$
$$CH=CH$$

(f) Presumably the pheromone detector has some very specific spatial requirement.

5. (a) $PCl_5/PCl_3/SOCl_2$ followed by KCN in ethanol.

(b) Electronegativities of C and N differ, so the C will be $\delta+$, the N, $\delta-$.

(c) $\delta-$ oxygen of water attacks $\delta+$ carbon.

(d) Pentanoic acid.

22. Ales and lagers

1. (a) Compounds with the same formula but different structures.

(b) $CH_3CH_2CH_2CH_2OH$, butan-1-ol; $CH_3CH_2CH(OH)CH_3$, butan-2-ol; $CH_3CH(CH_3)CH_2OH$, 2-methylpropan-1-ol; $(CH_3)_3COH$, 2-methylpropan-2-ol.

2. (a) The OH group is polar and hydrogen bonds with water molecules.

(b) The rest of the molecule is non-polar, so is insoluble in water. The bigger it is, the more it reduces the alcohol's water solubility.

(c) Solubility falls dramatically; there is now no —OH group to hydrogen bond.

3. (a) Any suitable oxidising agent: e.g. acidified potassium dichromate.

(b) $Cr_2O_7^{2-} + 8H^+ + 3C_2H_5OH \rightarrow 2Cr^{3+} + 3CH_3CHO + 7H_2O$

(c) Use excess ethanol. Do the reaction at 40–50°C, so the ethanol distils out.

(d) Ethanol reduces acidified dichromate to green Cr^{3+}. The amount of green Cr^{3+} is proportional to the amount of ethanol.

4. (a) Air oxidation.

(b) $2C_2H_5OH + O_2 \rightarrow 2CH_3CHO + 2H_2O$

(c) Oxidation to decan-2-one.

(d) Contains no primary or secondary alcohols, only tertiary or enols.

(e) Further oxidation of aldehydes to acids is easy. (Notice the high concentration of acids.)

5. (a) $C_2H_5OH + CH_3COOH \rightleftharpoons CH_3COOC_2H_5 + H_2O$

(b) Reflux. Use of acid catalyst.

6. Air oxidation of ethanol to ethanoic acid.

7. (a) 1042.

(b) Density and therefore OG raised by dissolved carbohydrates.

(c) Falls. Carbohydrates etc., replaced by less dense ethanol.

(d)

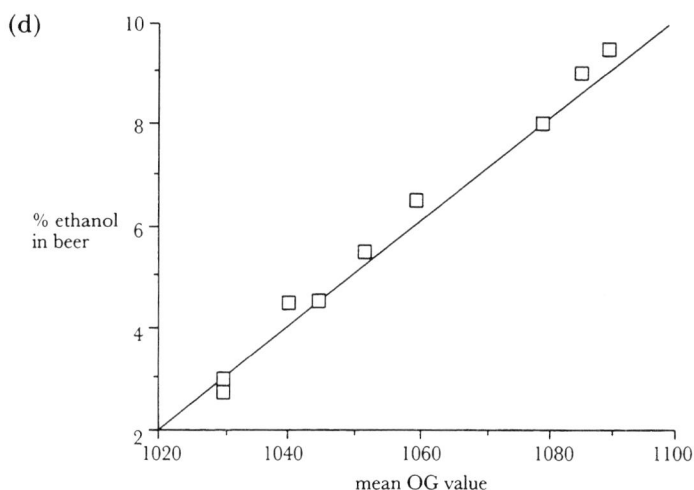

(e) 12.1% if graph is linear.

(f) To end up with 4.2% ethanol, we should have expected to start with an OG of about 1042, so an OG of 1030–1034 is very low. There will presumably be even less carbohydrate left after fermentation than usual.

(g) Duty is payable because it contains ethanol, and OG was only used because it was an easy estimate (albeit not perfectly accurate) of ethanol content. It is now EC policy to use % ethanol.

23. Toffee and toast

1. (a) One compound is converted into another of the same formula but different structure.

   (b) Joining of lots of identical molecules to make one long chain.

2. (a) Reagent with a spare pair of electrons that it will use to make a bond to another atom, usually a $\delta+$ or electron deficient one.

   (b) Polarised (carbon $\delta+$, oxygen $\delta-$), because of electronegativity difference between C and O.

3. (a) CN^-

   (b) $CH_3CHO + HCN \rightarrow CH_3CH(OH)CN$

   (c)

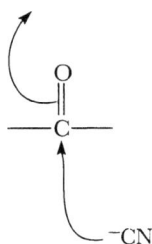

4. (a) The non-bonding pair on the OH group on the fifth carbon from the top.

   (b)

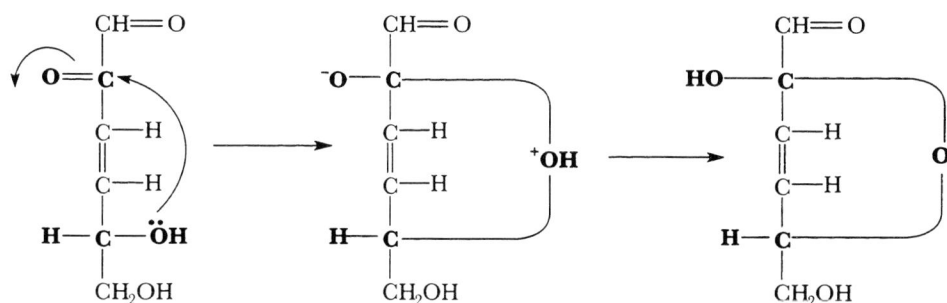

5. (a)

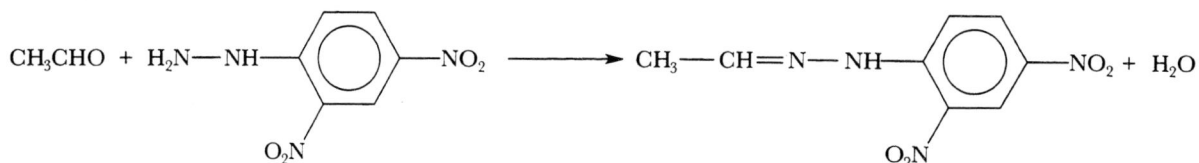

CH₃CHO + H₂N—NH— ... —NO₂ ⟶ CH₃—CH=N—NH— ... —NO₂ + H₂O

(b) The :NH₂– group.

(c)

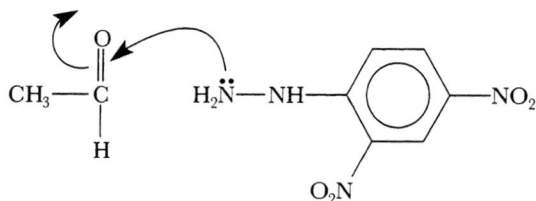

(d)

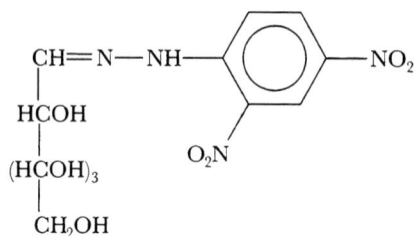

6. (a) RNH₂, or more specifically, the nitrogen atom.
   (b)

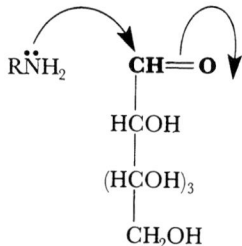

   (c) The lone pair of electrons is no longer available, occupied by the proton.
   (d) Not much vinegar is added, and it's a weak acid anyway.

7. Catalyst; it is regenerated at the end of box 5, so isn't actually used up.

8. Higher temperature accelerates browning reactions.

9.

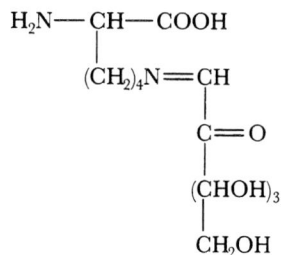

10. The Amadori compound has a C=O group in position 2 of the chain, which can form further Amadori products with other protein molecules.

Sand castles and mud huts © 1991 Jeffrey Hancock, published by Hodder and Stoughton Educationa

24. *Wrinkles and waves*

1. (a) Has a lone pair of electrons that it can use to form a bond.
   (b) The carbon carries a small + charge because of the electron withdrawing effect of the oxygen and chlorine atoms.
   (c)

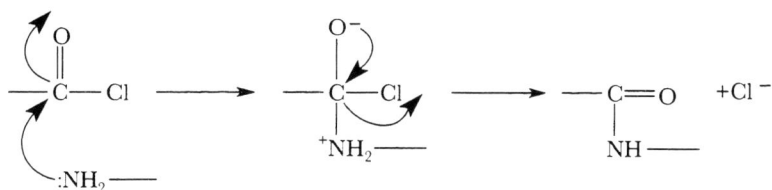

   (d)

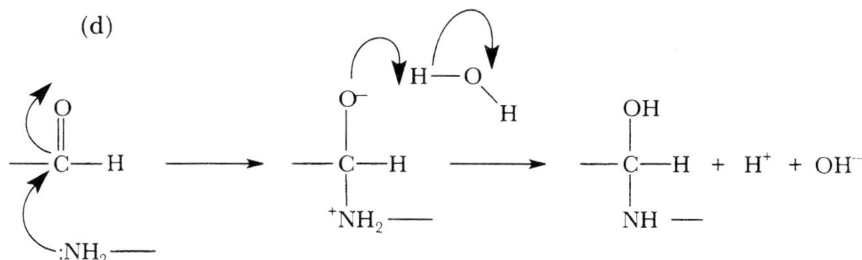

   (e) Cl^- is a good leaving group; H^- isn't. Cl is an electronegative atom; H isn't. The C—Cl bond is weaker than the C—H bond.
2. (a) Greater expense of producing it.
   (b) React the diacid with $PCl_5/PCl_3/SOCl_2$.
3. (a) Number of carbons in each monomer unit.
   (b) Nylon-4,6.
   (c) $-[CO(CH_2)_4CO-NH(CH_2)_{10}NH]-$
4. (a) Interaction between $\delta+$ hydrogen (bonded to O, N or F) and $\delta-$ O, N or F atom.
   (b) The H of the NH group interacts with the O of the CO group.
   (c)

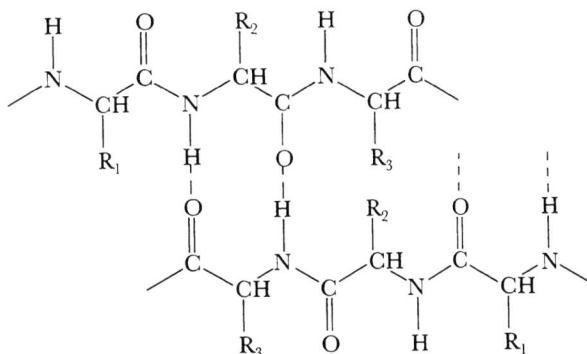

5. (a) Breakdown by water.
   (b) $-[CO(CH_2)_4CONH(CH_2)_6NH]- + H_2O \rightarrow$
      $-OC(CH_2)_4COOH + NH_2(CH_2)_6NH-$
   (c) Attack by lone pair of the oxygen of the water molecule on the carbon of the —CO— group.
   (d) OH^- is a better nucleophile.
   (e) Protonation of the —NH— group, making attack by water of the adjacent CO group easier.
6. (a) Hydrogen bonded to the —CO— or the —NH—.
   (b) Hair – i.e. proteins in general – have many more polar groups capable of H-bonding than nylon has. (This includes the amino acid side chains as well as the backbone.)
   (c) Water has been absorbed. Although it is still extensively hydrogen bonded, the structure is less rigid, as there is now

much more water incorporated in it. Why not weigh a piece of nail before and after immersion in hot water? If it absorbs water, it should gain mass.

(d) Removal of water by the hot iron, which hydrogen bonds in the desired flat state. Steam iron blows steam through the fabric to help realign the H bonds.

7. Interactions between polythene chains have only weak van der Waals' forces. Nylon is hydrogen bonded and therefore held more strongly.

8. Increasing a and b:
   (a) lowers water absorption: the polar —NH—CO— groups become less frequent.
   (b) lowers melting point; hydrogen bonding less frequent.
   (c) weakens it; less hydrogen bonding.
   (d) lowers it; hydrocarbon solvents interact best with hydrocarbon parts of the chains (or, to put it better, are excluded by the polar parts).

9. As far as I know, nobody knows. The tight curling indicates more hydrogen bonding, for some reason. That would also explain why it absorbs water less well, because there would be fewer polar groups not hydrogen bonded and thus available to H bond with water.

25. Shampoo and set

1. (a) (i) $-(CH_2)_4-NH_3^+$
       (ii) any one of the nitrogens protonated (most probably the $=NH$); more than one unlikely unless the pH is very low.
   (b) (i) $-CH_2-COO^-$
       (ii) $-CH_2-CH_2-COO^-$.
   (c) If the pH is above 7, the number of anions exceeds the number of cations.

2. (a) Become hydrated and separate from the rest of the surfactant molecule.
   (b) Like dissolves like, in this case using van der Waals' forces.
   (c) The $\delta+$ end of the water dipole interacts with the negative charge.
   (d) The sulphate has more electronegative atoms with which to interact with the water molecules.
   (e) The sulphate interacts with water molecules more strongly, so it dissolves in water better, dragging the rest of the molecule and the attached grease with it.

3. (a) At pH 6–7 hair carries a negative charge (question 1c), so the surfactant anion is repelled.
   (b) Negatively charged hairs repel each other and so fly up.

4. (a) Negatively charged hair attracts positively charged conditioner.
   (b) Conditioner attracted to hair, neutralises hair's negative charge.

5. (a) Thioglycolic acid converts —S—S— to —SH; not all the —S—S— bridges are reformed.
   (b) H_2O_2 oxidises some of the —SH to —SO$_3$H.
   (c) Fewer —S—S— bridges, which are partly responsible for hair's strength.

6. (a) Attack on —NH—CO— group, leading to hydrolysis of protein, forming amino acids. This will make the hair weaker, too.
   (b) (i) forms —CH$_2$—SO$_3$H.
       (ii) converts the primary alcohol to —CHO and —COOH.
       (iii) forms —CO—CH$_3$ from the secondary alcohol.

7. (a) Chance asymmetry in the electron distribution in one molecule or atom leads to a temporary polarity in the molecule. This in turn induces a polarity in an adjacent molecule, leading to attraction.

(b) Hydrogen bonding between the —OH groups of the polymer and the —CO—NH— groups of the protein.

(c) Polybut-2-ene: 4–6 kJ mol^{-1}: it has a slightly larger repeat unit than polyethene so will have slightly larger van der Waals' forces. Polyamide: 20–40 kJ mol^{-1}: more opportunities for hydrogen bonding than in polyethenol.

(d) Polyetheneimine is positively charged, so can interact strongly with negatively charged groups in the hair.

26. Flashes and bangs

1. (a) $4C_7H_5N_3O_6 + 21O_2 \rightarrow 28CO_2 + 10H_2O + 6N_2$
    (b) 3.71 moles of carbon.
    (c) Black smoke.
2. (a) Concentrated nitric and sulphuric acids.
    (b) NO_2^+.
    (c) Electrophile.
    (d) $H_2SO_4 + HNO_3 \rightarrow H_2NO_3^+ + HSO_4^-$, then $H_2NO_3^+ \rightarrow H_2O + NO_2^+$.
    (e)

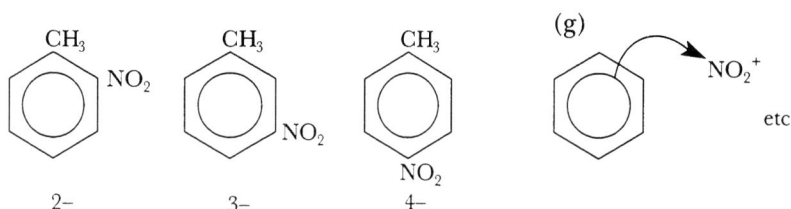

2–   3–   4–   (g) etc

(f) 2 – and 4 – isomers formed.

(h) Electron withdrawing effect of the NO_2 group makes the electron density of the ring much less, thus less susceptible to attack by electrophile.

3. (a) Harder. Chlorine atoms are electron-withdrawing.
    (b) The π cloud of the aromatic ring repels nucleophiles away from the ring. (And p–π overlap between the Cl atom and the ring strengthens the ring-Cl bond.)
    (c) The combined electron withdrawing effect of all six groups/atoms.
4. (a) Easier.
    (b) Overlap between oxygen lone pair and π cloud of ring feeds electron density onto the ring.
    (c) $2HOC_6H_2(NO_2)_3 + PbO \rightarrow Pb[OC_6H_2(NO_2)_3]_2 + H_2O$
    (d) Phenols aren't as strong acids as CO_2/H_2CO_3, so can't displace CO_2/H_2CO_3 from salts.
    (e) Electron withdrawing effect of three NO_2 groups makes ionisation of the —OH group easier, thus increasing acid strength.
5. (a) Sn + concentrated HCl.
    (b) Limited contact time with reagent.
    (c) $NaNO_2$ + dilute HCl, producing HNO_2 in the reaction flask. Temperature below 10°C.
    (d)

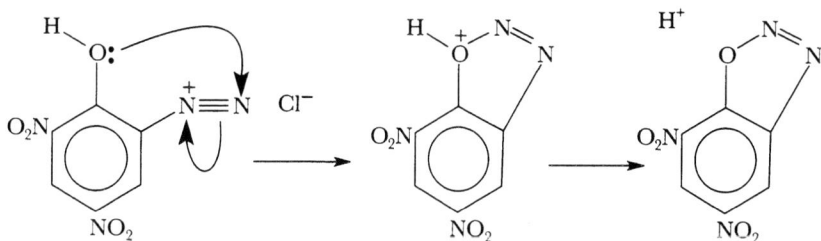

27. Spare parts

1. (a) Without animal testing, we have no knowledge of its possible toxicity. An animal's life is not as significant as that of a human.
   (b) All life is sacred; what right have we to kill anything? Animal testing only gives a guide to toxicity; the human metabolism is different, so human testing will ultimately be needed.
   (c),(d) I'm not going to air my prejudices!
2. (a) 90% of the PTFE molecules are arranged regularly as in a crystal.
   (b) Crystals are regular and rigidly bonded; glasses are rigid but irregular.
   (c) Chains cannot slide relative to one another.
   (d) The polymer chains pack together badly and are only weakly bonded together.
3. (a) Heat.
   (b) The free radical attacks another alkene, generating another free radical:

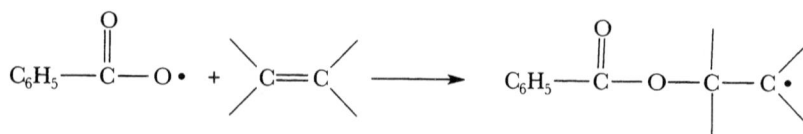

$$C_6H_5-\overset{\overset{\displaystyle O}{\|}}{C}-O\bullet \quad + \quad \underset{/}{\overset{\backslash}{C}}=\underset{\backslash}{\overset{/}{C}} \quad \longrightarrow \quad C_6H_5-\overset{\overset{\displaystyle O}{\|}}{C}-O-\overset{|}{\underset{|}{C}}-\overset{/}{\underset{\backslash}{C}}\bullet$$

and so on.

   (c) Two radicals reacting together.
   (d) Too much initiator gives a high concentration of radicals, which will tend to cause frequent termination and hence short chains.
   (e) The light provides the energy to break the π (or other) bonds, thus creating radicals. These then react as before.
4. (a) Lots of polar groups to hydrogen bond to the water molecules.
   (b) HEMA; more polar groups to H bond.
   (c) Awkward shaped, so poorly packed, thus "porous" on a molecular scale. And the polymer contains much water, in which oxygen is slightly soluble.
5. (a) $HOCH_2-CH_2OH + HOOC-C_6H_4-COOH \rightarrow -[OCH_2-CH_2O-OC-C_6H_4-CO]- + H_2O$
   Warm ethane-1,2-diol and benzene-1,4-dicarboxylic acid together, with a small amount of an acid catalyst, such as H_2SO_4.
   (b) Hydrolysis of the ester, converting it back to the parent compounds, catalysed by acid or base.
   (c) pH of body fluids generally 7–7.6, so hydrolysis very slow.
6. (a) Only one Cl per molecule, so only one OH formed, so only the dimer, $(CH_3)_3Si-O-Si(CH_3)_3$, is formed.
   (b) Every time a trimethyl compound goes into the chain, it stops it.
   (c) CH_3SiCl_3 produces $CH_3Si(OH)_3$, which can form three Si—O—Si bridges.
   (d) Three links inevitably cross link chains.
   (e) The more cross linking, the firmer the silicone.

28. Headache or hangover?

1. (a) 2-hydroxybenzoic acid or 2-hydroxybenzenecarboxylic acid.
   (b) Boil bark with water, filter off bark debris, cool to crystallise, filter.
2. (a) (i) Soluble because of polar groups: —COOH and —OH.
       (ii) Not very soluble because of insoluble aromatic ring.
   (b) Charged species, like the aspirin anion, interact more strongly with water.

3. (a) Hydrolysis.

$HOOC(C_6H_4)OCOCH_3 + H_2O \rightarrow HOOC(C_6H_4)OH + CH_3COOH$

(b) OH^- is a better nucleophile than H_2O.

$HOOC(C_6H_4)OCOCH_3 + 3NaOH \rightarrow NaOOC(C_6H_4)ONa + CH_3COONa + 3H_2O$

(c) $2HOOC(C_6H_4)OCOCH_3 + CaCO_3 \rightarrow Ca^{2+}(^-OOC(C_6H_4)OCOCH_3)_2 + CO_2 + H_2O$

(d) Aq $Ca(OH)_2$ could hydrolyse the aspirin, just like NaOH.

(e) It is converted back to aspirin, and some hydrolysis starts.

4. (a) Salicylic acid, paracetamol.

(b) Aspirin.

(c) Paracetamol.

(d) Ibuprofen. (The commercial drug is presumably the racemic mixture. Whether there is any difference between the two isomers, as there was with thalidomide and now seems to be the case with the asthma drug salbutamol, I don't know.)

(e) Paracetamol.

5. —COOH groups are weak acids; the stomach contains HCl, which is much stronger. Anyway, ibuprofen also contains —COOH, but is much less irritating. (The irritation is now thought to be a prostaglandin effect, too.)

6. 40 mg switches off the thromboxane A_2, without affecting prostacyclin synthesis.

7. (a) A: Conc nitric and conc sulphuric acids, room temperature.

B: Oxidation with e.g. alkaline $KMnO_4$, refluxed, followed by acidification with dil H_2SO_4.

C: Sn + conc HCl, followed by excess aq NaOH.

D: $NaNO_2$ + dil HCl, warmed.

E: CH_3COCl or $(CH_3CO)_2O$.

(b) (i) Use $\mathbf{CH_3COCl}$ or $\mathbf{(CH_3CO)_2O}$.

(ii) Oxidise to $\mathbf{CH_3COOH}$ (acidified $K_2Cr_2O_7$, refluxed), then add PCl_5 or $SOCl_2$ to form $\mathbf{CH_3COCl}$. Then as above.

8. (a) A: $CH_3CH=CH_2$, with $AlCl_3$ or H_3PO_4 as catalyst.

B: Air or O_2.

C: Warm dilute H_2SO_4.

(b) Propanone.

(c) An acid; e.g. dilute H_2SO_4.

(d)

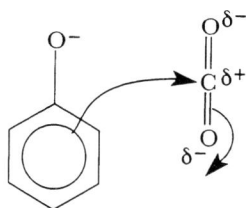

(e) The anion is a better nucleophile than the parent phenol because it is more electron-rich.

(f) Electrophilic substitution of CO_2 (step D) gives 2— and 4— isomers.

(g) The 4— isomer has intermolecular H-bonding, which makes it less volatile than 2—, which has intramolecular H-bonding.

29. Impossible – yet perhaps it works?

This is, inevitably, more open to discussion, but I have tried to bring out some points about the nature of science.

1. (a) Chromatography. Make separate solutions of "Painomin" and each of the pure drugs in some suitable solvent. Spot onto chromatography plate (tlc works; I don't think paper does,

although that is immaterial for this question), dip plate into some solvent. Locate spots with some suitable reagent. "Painomin" should give two spots which have risen to the same height as paracetamol and codeine. The aspirin spot should not match.

(b) It is possible that two substances were present that behaved the same way as paracetamol and codeine under the conditions of the experiment. (Of course, we can be quite certain that there is no aspirin. This relates to Popper's ideas: disproof of a theory is incontrovertible.)

(c) Repeat with different eluting solvents. Very unlikely (but never impossible) that different substances would behave the same under different conditions.

2. (a) Get patients to assess their own pain (no-one else can!), for example on a 3 point scale. (0 = no better; 1 = some improvement; 2 = major improvement/no pain.)

(b) (i) Psychological factors are known to affect health. If patients expect to get better, they often do.
(ii) Do not let patient know whether tablet is drug or not.

(c) (i) To avoid fraud (the doctor might try to adjust the results) and the placebo effect. If doctor knows which drug is which, the patient might find out by some change in the doctor's manner towards him.
(ii) Code the drugs by letters or numbers, so that no-one involved in the trial knows what is what. Someone else holds the key to the code; break the code only after the trial is over.

(d) Trial requires that some patients receive placebo: i.e. no drug at all, and some of these will thus experience pain that could be relieved better by drugs.

3. 1 in 10^{60}.

4. (a) 6×10^{23} molecules of drug per dm³.
(b) $6 \times 10^{23}/10^{60}$ molecules; i.e. about 10^{-37}, effectively zero, molecules per dm³.

5. No modifications necessary.

6. Each of these views probably has its supporters. Some comments:

(a) It is difficult to dismiss the evidence. Apart from the hundreds of doctors (and vets) who practise homeopathy because in their experience it works, one cannot discount the proper double blind studies that have been done. Of the five I have seen in *reputable* scientific journals, 3 found a significant effect, 2 found none.

(b) I cannot see that a trial considered adequate for, say, aspirin, suddenly ceases to be adequate if it deals with homeopathic remedies. An editorial in the "Lancet" has made the same point.

Sand castles and mud huts © 1991 Jeffrey Hancock, published by Hodder and Stoughton Educational